Cardiovascular Diseases

Cardiovascular Diseases

Genetic Susceptibility, Environmental Factors and their Interaction

Edited by

Nikolaos Papageorgiou, MD, PhD

ELSEVIER

AMSTERDAM • BOSTON • HEIDELBERG • LONDON
NEW YORK • OXFORD • PARIS • SAN DIEGO
SAN FRANCISCO • SINGAPORE • SYDNEY • TOKYO

Academic Press is an imprint of Elsevier

Academic Press is an imprint of Elsevier
125 London Wall, London EC2Y 5AS, United Kingdom
525 B Street, Suite 1800, San Diego, CA 92101-4495, United States
50 Hampshire Street, 5th Floor, Cambridge, MA 02139, United States
The Boulevard, Langford Lane, Kidlington, Oxford OX5 1GB, United Kingdom

Library of Congress Cataloging-in-Publication Data
A catalog record for this book is available from the Library of Congress

British Library Cataloguing-in-Publication Data
A catalogue record for this book is available from the British Library

ISBN: 978-0-12-803312-8

Publisher: Mica Haley
Acquisition Editor: Stacy Masucci
Editorial Project Manager: Samuel Young
Production Project Manager: Edward Taylor
Designer: Matthew Limbert

Typeset by TNQ Books and Journals

Dedication

To my mentor and friend,
Professor Dimitris Tousoulis

Contents

List of Contributors

E. Androulakis John Radcliffe Hospital, Oxford, United Kingdom

A. Briasoulis Wayne State University/Detroit Medical Center, Detroit, MI, United States

J. Chandrasekhar Icahn School of Medicine at Mount Sinai, New York, NY, United States

M. Charakida University College London, London, United Kingdom

S. Chiesa University College London, London, United Kingdom

T. Christophides Barts Heart Centre, Barts Health NHS Trust, London, United Kingdom

R.M. Dumitru St Peter's Hospital, Chertsey, Surrey, United Kingdom

E. Farmaki 1st Cardiology Department, Hippokration Hospital, Athens University Medical School, Athens, Greece

P. Glennon University Hospital Coventry, Coventry, United Kingdom

G. Hatzis 1st Cardiology Department, Hippokration Hospital, Athens University Medical School, Athens, Greece; Philipps University Marburg, Germany

S.A. Hayat University Hospital Coventry, Coventry, United Kingdom

K. Karatolios Philipps University Marburg, Germany

P.B. Lim Imperial College NHS Health Care Trust, London, United Kingdom

C.J. McAloon University Hospital Coventry, Coventry, United Kingdom

R. Mehran Icahn School of Medicine at Mount Sinai, New York, NY, United States

M. Michail Barts Heart Centre, Barts Health NHS Foundation Trust, London, United Kingdom

K. Norrington Barts Heart Centre, Barts Health NHS Trust, London, United Kingdom

E. Oikonomou 1st Cardiology Department, Hippokration Hospital, Athens University Medical School, Athens, Greece

F. Osman University Hospital Coventry, Coventry, United Kingdom

N. Papageorgiou Barts Heart Centre, St. Bartholomew's Hospital, London, United Kingdom; Athens University Medical School, Athens, Greece

G. Siasos 1st Cardiology Department, Hippokration Hospital, Athens University Medical School, Athens, Greece

D.I. Swerdlow University College London, London, United Kingdom; Imperial College London, London, United Kingdom

D. Tousoulis 1st Cardiology Department, Hippokration Hospital, Athens University Medical School, Athens, Greece

E. Zacharia 1st Cardiology Department, Hippokration Hospital, Athens University Medical School, Athens, Greece

K. Zacharias Croydon University Hospital, London, United Kingdom

M. Zaromitidou 1st Cardiology Department, Hippokration Hospital, Athens University Medical School, Athens, Greece

Acknowledgments

I would like to record my most sincere thanks to all the people from Elsevier who have contributed to the production of this book.

All the chapter contributors deserve special thanks and are listed on the page above.

Chapter 1

Summary

N. Papageorgiou[1], D. Tousoulis[2]
[1]St. Bartholomew's Hospital, London, United Kingdom; [2]Athens University Medical School, Athens, Greece

Cardiovascular disease (CVD) is the leading cause of mortality and morbidity in modern societies and encompasses a range of human pathology. It is highly suspected that CVD is strongly related to genetic and environmental factors, while studies have demonstrated that there is an interaction between genetic and environment factors that can lead to CVD.

In this book, we aim to review the current knowledge on this topic:

Zaromitidou et al. discuss thoroughly the role of atherosclerosis as well as the role of genetics in CVD, providing related mechanistic links. Further to this, Chandrasekhar et al. review the current data on the impact of genetic variations on antiplatelet therapy, as well as the potential role of genotyping in prescription of antiplatelet therapies in acute coronary syndromes. In addition, MacAloon et al. examine the current cardiovascular epidemiology and perform specific analyses on fatal and nonfatal CVD burden. They also examine risk factor exposure and prevalence to determine explanations for the current pattern of CVD. Specific CVD categories and conditions are also examined to elucidate a more definitive analysis of global CVD determinants. Moreover, Chiesa et al. review the current evidence surrounding the impact of lifestyle choices on cardiovascular risk factors and disease, and they discuss the beneficial effects of modifying these behaviors with regard to morbidity and mortality. Hatzis et al. summarize the thus-far-acquired knowledge on the main pollutants and present the basic mechanisms that mediate their actions on the vasculature.

Briasoulis et al. acknowledge the contribution of candidate gene studies to complex CVDs such as acute coronary syndromes, and the fact that the clinical impact and the pathophysiological implications of polymorphisms are elusive. On top of these data, Swerdlow et al. discuss the different relationships that a risk factor can have with disease and the role of genetic studies in causal inference, and they evaluate the insights provided by genetics into the contribution of some prominent risk factors to CVD pathogenesis.

Furthermore, Dumitru et al. provide a few examples of genetic susceptibility in biochemical and physiological traits of CVD. They also examine the opportunities

Cardiovascular Diseases.

provided by pharmacogenomics for future improvements in treatment of CVD. Finally, Norrington et al. introduce key concepts in gene–environment interactions in CVD. The expounding importance of gene–environment interactions in understanding the missing heritability of multifactorial CVDs is highlighted as well as the fundamental concepts of statistical and biological interactions.

The available data are promising, and there are still ongoing studies aiming to evaluate the interactions between genetics and environmental factors in CVD. We hope that the information and data provided will be interesting for the readers of the book and will stimulate further research on the topic.

Chapter 2

Atherosclerosis and Coronary Artery Disease: From Basics to Genetics

M. Zaromitidou, G. Siasos, N. Papageorgiou, E. Oikonomou, D. Tousoulis
Athens University Medical School, Athens, Greece

INTRODUCTION

Cardiovascular disease is the leading global cause of death. According to the World Health Organization, 17.3 million deaths resulted from cardiovascular disease worldwide in 2008, and that number is expected to reach 23.3 million by 2030. The underlying pathology of cardiovascular disease is atherosclerosis and, depending on the artery affected, can manifest as CAD or as cerebrovascular or peripheral disease. Coronary artery disease (CAD) is responsible for 7.3 million deaths and is the second cause of death in people under the age of 59. It is also noteworthy that CAD is listed among the first causes for years of life lost due to premature death or disability, and thus has important social and economic impact. Low- and middle-income countries exhibit a continuous increase in the prevalence of CAD in contrast to developed countries. This discrepancy is attributed to the ineffective primary and secondary prevention measures in addition to poor healthcare systems in the low- and middle-income countries [1].

The mechanisms of atherosclerosis have gained increasing attention over recent decades. Many questions still remain concerning the heterogeneity that atherosclerosis displays in time (when?), in the areas affected (where?), in the factors triggering the initiation (how?), and in defining the natural history and evolution of the disease (why?). Atherosclerosis is a chronic systematic disease involving large and medium-sized arteries that initiates early in life [2]. As it progresses, the structure of normal arteries is modified, and atherosclerotic plaques are formed with consequent narrowing or dilation of the vessel. Atherosclerosis is a systematic disease with local manifestations, but the factors determining the preferential involvement of a vascular region over another (ie, coronary or carotid artery) as well as the development of atherosclerotic lesions in certain sites of a specific artery (eg, branches, curvatures, and proximal areas

Cardiovascular Diseases.

of left anterior descending) are unclear. In addition, regardless of tremendous research efforts, sparse data are available illuminating the factors responsible for the transition of a stable atherosclerotic plaque to a vulnerable plaque that can cause the acute complications of atherosclerosis (eg, myocardial infarction (MI) in CAD). Diversity also exists in the biological procedures and their clinical manifestations following rupture or erosion of unstable plaques (clinically silent MI, unstable angina, and MI with or without ST elevation).

The biggest question to be answered is why the human arteries are vulnerable to atherosclerotic changes. As most communicable diseases are now successfully treated and average life expectancy has increased, cardiovascular disease burden emerges as one of the most important health issues. It was some decades ago that the Framingham Heart Study provided valuable data regarding the primary prevention of atherosclerotic cardiovascular disease and established the *cardiovascular risk factor* as a new term. The identification of several risk factors associated with modern lifestyle such as hyperlipidemia, smoking, hypertension, obesity, diabetes mellitus, lack of exercise, anxiety, and depression supports the notion that atherosclerosis is a disease of urbanization [3]. However, well-known risk factors such as older age, family history of cardiovascular disease, male sex, and genetic abnormalities (familial hypercholesterolemia) indicate a genetic contribution to atherosclerosis. The report of atherosclerotic findings in mummies from populations of disparate regions with differences in dietary habits and certainly differences in dietary and lifestyle patterns compared to the present time suggests that human vessels are susceptible to atherosclerotic alterations regardless of the current environmental risk factors. Moreover, these observations underline the ongoing need for novel risk factors [4]. Therefore, both genetic and environmental factors are implicated in the pathogenesis of atherosclerosis, whereas their interaction may account for the heterogeneity that atherosclerosis displays.

NORMAL ARTERY STRUCTURES

The normal artery wall comprises three layers: the tunica intima, the tunica media, and the tunica adventitia.

- The tunica intima consists of the endothelium, connective tissue (collagen, laminin, fibronectin, and other extracellular matrix molecules), and a basal layer of elastic tissue called internal elastic lamina that separates the tunica intima from tunica media. Endothelium is a thin monolayer of cells that serves as the contact surface with blood. Due to its strategic location, the endothelium has emerged as the main regulator of vascular homeostasis with its structural and functional properties altered in response to local and systemic stimuli [5].
- The tunica media is characterized by the presence of concentric layers of vascular smooth muscle cells (VSMCs) and elastin-rich extracellular matrix. It is the final recipient of signals regulating the vascular tone and is separated from adventitia by the external elastic lamina.

- The tunica adventitia is the outer layer of the vascular wall, and it consists of fibroblasts, collagen, mast cells, nerve endings, and vasa vasorum. It was not until recently that important functions of the adventitia were identified. More specifically, adventitia seems to participate in the cell trafficking through the arterial wall and the signaling between vascular endothelial cells, smooth muscle cells, and the local tissue environment. In addition, this layer is involved in the repair mechanism following vessel injury, in the regulation of the dynamic lumen size (via medial smooth muscle tone), and in the inward or outward wall-remodeling response [6].

PATHOGENESIS OF ATHEROSCLEROSIS

Atherosclerosis is a complex disease with pieces of the pathophysiology puzzle still missing. In order to prevent or treat atherosclerotic complications, it is imperative to clarify and comprehend the mechanisms involved in the pathogenesis. Endothelial cells, VSMCs, and arterial extracellular matrix macromolecules (collagen, proteoglycans, and elastin) are components of a normal artery that play a crucial role in the atherosclerosis process. Lipid accumulation, leukocyte recruitment, and local flow hemodynamic and inflammatory mechanisms are essential elements of the underlying pathology, whereas plaque angiogenesis and mineralization contribute to the evolution of atheroma.

In brief, altered endothelial function leads to transportation and subendothelium accumulation of low-density lipoprotein (LDL) particles. Retention of LDL results in its modification to oxidized LDL (Ox-LDL), triggering an inflammatory cascade and activating endothelial cells to signal the recruitment and migration of monocytes. The coexistence of monocytes and lipids in the subendothelium leads to the phagocytosis of the latter by monocytes and the formation of foam cells that gradually progress into atherosclerotic lesions (Fig. 2.1).

INITIATION OF ATHEROMA

The Role of Endothelium

The endothelium plays a crucial part in the initiation and in all stages of atherosclerosis due to its role as the barrier between blood flow and arterial layers. Normal endothelial cells sense the changes in the microenvironment and regulate important functions such as vascular tone, circulating cell adhesion, coagulation, fibrinolysis, vessel wall inflammation, and response to hemodynamic changes. In addition, endothelial cells are arranged in a specific way, forming a tight seal and thus controlling the transportation of all molecules between the lumen and the intima.

In order for atherosclerosis to initiate, LDL needs to cross the endothelium and reside in the subendothelium. Increased endothelium permeability is part

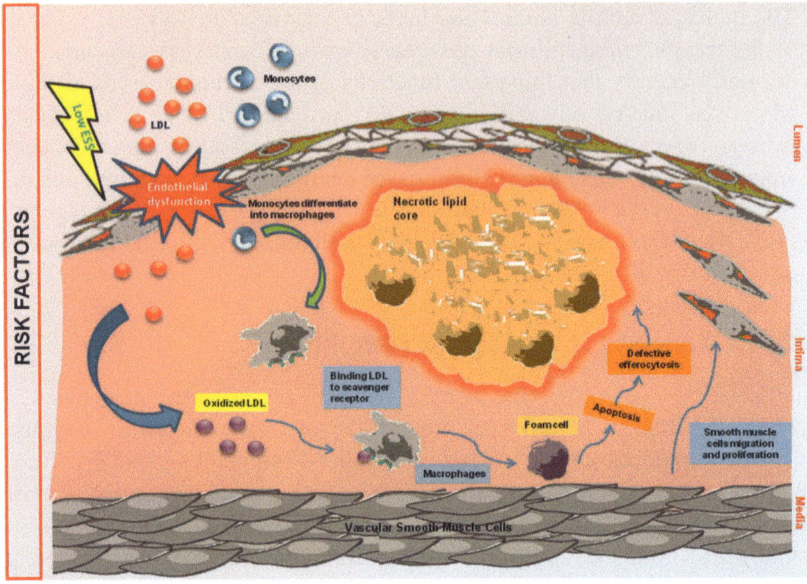

FIGURE 2.1 Low shear stress (ESS) promotes endothelial dysfunction, transportation, and sub-endothelium accumulation of low-density-lipoprotein (LDL) particles. Retention of LDL results in its modification to oxidized LDL, which triggers an inflammatory cascade that activates endothelial cells and signals the recruitment and migration of monocytes. After crossing the endothelium, monocytes differentiate into macrophages, internalize modified LDL, and form foam cells. The infiltration of macrophages within the lipid pool and their subsequent apoptosis result in the development of the lipid-rich necrotic core that combined with a collagen fibrous cap constitute the advanced lesion called fibroatheroma.

of endothelial dysfunction and a precondition for LDL accumulation. Several factors can impair normal endothelial function, including smoking, hyperlipidemia, arterial hypertension, and diabetes mellitus. It is noteworthy that whereas these factors affect the whole artery, atherosclerosis occurs only in certain vessel areas (curved regions, branches, and bifurcation intersections) that are mainly characterized by disturbed flow patterns. On the contrary, atherosclerotic lesions develop rarely in arterial regions with accordingly few branches (internal mammary artery) and laminar blood flow. Predilection sites are complicated with atherosclerosis in the presence of cardiovascular risk factors.

Mechanosensors positioned on the cell surface identify endothelial shear stress (ESS) stimuli and activate atheroprotective or atheroprone intracellular pathways. Low ESS promotes an atheroprone endothelial cell phenotype (endothelial activation) via up- or down-regulation of gene expression. Regarding the initiation of atherosclerosis, low ESS upregulates the expression of genes encoding for the LDL receptor of the endothelial membrane. In addition, low ESS contributes to the increased endothelial LDL permeability by inducing structural alterations to the cells that result in the transition from fusiform to polygonal shape and widening of the junctions between the cells. The attenuation of mitotic and endothelial apoptotic cell circles, in accordance with the prolonged residence time of LDL near the endothelium due to low ESS, contributes to the augmented membrane permeability [7]. Thus, in the presence of hyperlipidemia, LDL particles cross the endothelium and start to accumulate in the intima tunica.

Endothelial activation promotes early atherosclerosis not only by affecting extracellular LDL accumulation but also by diminishing the atheroprotective properties of the endothelial-derived nitric oxide (NO). Specifically, NO synthesized in the normal endothelial cells modulates vascular tone (favors vasodilation), hampers leukocyte migration to the intima by reducing the expression of adhesion molecules, and decreases VSMC proliferation. Whereas laminar blood flow upregulates the genes implicated in NO production, disturbed flow and low ESS act in an opposite manner.

Moreover, activated endothelial cells promote the atherosclerotic process by increasing the production of reactive oxygen species (ROS) and the secretion of cytokines and chemokines that trigger the migration and activation of monocytes. An additional atherogenic characteristic of endothelial activation is the release of growth factors such as platelet-derived growth factor (PDGF) that induce VSMC proliferation and extracellular matrix synthesizing. The accumulation of VSMCs and extracellular molecules, especially proteoglycans, is known as *intimal thickening* or *intimal hyperplasia* [8].

The Role of LDL Particles

The accumulation of cholesterol in the subendothelium is an essential step for the initiation of atherosclerosis. LDL particles transfer cholesterol through blood circulation and into the arterial intima. The endothelial cell membrane is considered impermeable and highly controls molecule trafficking between the blood flow and the vessel wall. In sites of endothelial dysfunction, LDL can penetrate through the cells and accumulate in the intima. The negative charged proteoglycans of extracellular matrix (ECM) bind LDL, resulting in the retention of the lipoproteins – a key step for the initiation of atherosclerosis. The LDL–proteoglycan interaction entraps LDLs in the subendothelium and prolongs their residence time. Thus, LDLs are susceptible to oxidative modification generated by ROS and a number of enzymes such as nitric oxide synthases, NADPH oxidases, lipoxygenases, and myeloperoxidases [9]. In addition, modified LDL aggregates are produced by lipases and proteases such as secretory phospholipase A2 (sPLA2) and sphingomyelinase under proteoglycan regulation. LDL particles may also be chemically modified by nonenzymatic glycation in patients with chronic hyperglycemia, underlining diabetes mellitus as a major risk factor for cardiovascular disease [10].

Ox-LDL displays pro-atherogenic properties that aggravate atherosclerotic pathology. In particular, Ox-LDL induces tissue factor expression and platelet aggregation, limits the beneficial biological actions of NO, and promotes endothelial damage and apoptosis. Ox-LDL also affects the genetic substrate by upregulating genes associated with inflammation. Chemotactic molecules such as monocyte chemotactic protein-1 (MCP1) and PDGF are secreted, inducing the recruitment and proliferation of monocytes and VSMCs. Macrophages express scavenger membrane receptors for Ox-LDL, whereas VSMCs increase collagen ECM concentration. Overall, Ox-LDL further amplifies endothelial activation, inflammation, and thrombogenicity.

The Role of Inflammation

Monocyte-Derived Macrophages

Modified LDL is identified as a pathogen, and thus an inflammatory response is initiated. The first step in the inflammatory process is the recruitment of leukocytes, mostly monocytes. Endothelial cells with normal function are not susceptible to transmigration of monocytes. The activated endothelial cells and modified LDL reverse the endothelium microenvironment in favor of monocyte adherence and migration by triggering the expression of adhesion molecules on endothelial cell membrane and the secretion of chemoattractant cytokines. Molecules known to be involved in monocyte adhesion are the vascular cell adhesion molecule-1 (VCAM1), the intracellular adhesion molecule-1 (ICAM1), and members of the selectin family (P- and E-). In order for adherent monocytes to penetrate the endothelium, the secretion of chemokines must precede. MCP1 and fractalkine are the main chemokines implicated in the transmigration of monocytes through the endothelial layer [11].

Once monocytes overcome the endothelium barrier and reside in the intima, they differentiate into macrophages. Macrophages in the atherosclerotic lesion present marked phenotype heterogeneity that is defined by the environmental stimuli. Their categorization is based on the surface markers they express, the molecules they secrete, and their biological functions. The most abundant phenotype in the subendothelium is the M1 macrophage, which is induced by a variety of stimuli such as proinflammatory cytokines (eg, interferon-γ (IFNγ) and tumor necrosis factor (TNF)), cholesterol crystals and esters, and modified LDL. M1 macrophages play an essential role in the immune response against pathogens during an infection mediating the production of ROS. In atherosclerotic lesions, modified LDL induces chronic activation of M1 macrophages, resulting in tissue damage. Cholesterol crystals and esters promote the M1 phenotype by triggering the caspase-1 activating NLRP3 inflammasome, the Toll-like receptor-4 (TLR4) or nuclear factor–kappa B (NFκB)-mediated pathway. Oxidized LDL leads to M1 differentiation by hampering Kruppel-like factor-2 and stimulating the TLR4 pathway [12]. Macrophages expressing the M1 phenotype are characterized as proinflammatory and secrete a number of cytokines, such as interleukin-6 (IL6), IL1β, and TNF that further aggravate atherosclerosis [13]. The M2 macrophage is a different phenotype that is induced by stimuli such as IL4 and IL10 and displays antiinflammatory properties, including the clearance of apoptotic cells [14]. Interestingly, the phenotype of monocytes depends on environmental stimuli that in turn can be modulated by the activated macrophages. In addition, macrophages can switch between different phenotypes in response to environmental changes, a function known as *plasticity* [15].

Monocyte-derived macrophages internalize the LDL particles, forming the so-called foam cells. Although macrophages express LDL receptors, foam cell formation is not mediated through them, as they are downregulated early by the accumulated cholesterol levels. Modified LDL uptake is mediated by scavenger receptors, whereas aggregated LDLs and native LDLs are engulfed by phagocytosis and pinocytosis mechanisms, respectively. The scavenger receptors participating in foam cell formation are scavenger receptor-A (SRA), CD36, scavenger receptor class B member 1 (SRB1), and lectin-type receptor-1 (LOX1), with the first two internalizing the majority of modified LDLs [16]. Once lipoproteins enter the macrophage, they are transferred to the late endosome/lysosome, where they are subjected to hydrolysis by lysosomal acid lipase and form free cholesterol. Subsequently, free cholesterol is re-esterified in the endoplasmic reticulum (ER) by cholesterol acyltransferase-1 (ACAT1) to form cholesteryl fatty acid esters, and it is eventually stored in lipid droplets [17]. The re-esterification in the ER is an essential and prophylactic step against the cytotoxicity of free cholesterol [18]. Neutral cholesterol ester hydrolase transforms cholesterol esters into free cholesterol that can efflux macrophages through ABCA1 and ABCG1 transporters, and thus it comprises a crucial step in the initiation of reverse cholesterol transport. The accumulation of cholesteryl esters in the macrophage due to hypercholesterolemia ultimately provides a foamy appearance (foam cells).

LDL and Inflammation

The internalization of lipoproteins further aggravates endothelial dysfunction and inflammation. Modified LDLs function as a ligand for the TLR signaling pathways that mediate the release of proinflammatory cytokines and chemokines. In particular, the combined activation of TRL4, TRL6, and CD36 by Ox-LDL triggers the NFκB pathway and the expression of chemokines that amplify monocyte recruitment. Ox-LDL, oxidized phospholipids, and fatty acids when in a prolonged ER stress environment trigger TRL2, TRL6, and CD36 into promoting macrophage apoptosis. However, the integration of free cholesterol in the macrophage membrane can also initiate TRL activation. Furthermore, lipoproteins contribute to inflammation and atherogenesis through the activation of inflammasome. Specifically, this pathway is induced by cholesterol crystals resulting in the secretion of IL1β and IL18 [19].

Atheroprotective Mechanisms and Inflammation

Macrophages exhibit certain mechanisms to counterbalance lipoprotein accumulation and inflammation and maintain homeostasis. Oxysterol and desmosterol, intermediate molecules in cholesterol biosynthesis, induce the activation of liver X-receptor-mediated antiinflammatory response (LXR). In particular, LXR upregulates the expression of ABCG1 and ABCA1 transporters that efflux cholesterol onto high-density lipoprotein (HDL) particles and apolipoprotein A1, respectively, in order to deliver it back to the liver. There seems to be an interaction between LXR activity and the TLR pathway, as the former reduces the TRL stimulus (cholesterol) whereas TRL activation downregulates the LXR expression. Another example of the delicate balance between inflammatory and antiinflammatory mechanisms is the paradoxical amplification of the atherosclerotic process in genetically deprived NFκB mice, attributed to parallel suppression of IL10 [11,20].

Retention of Macrophages

The progression of atherosclerosis is interconnected with the number of resident macrophages. The majority of macrophages in atherosclerotic lesions was until recently thought to originate from the recruitment and transmigration of the circulatory reservoir. According to new findings, resident macrophages mostly originate from the proliferation occurring after lipid internalization and macrophage activation. The exact mechanism involved in macrophage proliferation is not fully elucidated, but modified LDL uptake via scavenger receptor A seems to play a key role [21]. Moreover, the accumulation of cholesterol promotes macrophage retention via the expression of netrin-1 and semaphoring 3E proteins, thus regulating the number of macrophages populating the lesion [22,23]. However, the number of macrophages decreases due to emigration and apoptosis. Macrophages emigrate through the media to the adventitia or directly to the lumen under stimuli that are yet unknown. Apoptosis is a significant procedure in atherosclerosis evolution and is analyzed in this chapter.

T Lymphocytes

Inflammatory mechanisms in atherosclerosis involve not only monocytes but also T lymphocytes, B lymphocytes, and dendritic and mast cells. The proinflammatory markers released by macrophages trigger T-lymphocyte transmigration into the intima, where the latter secrete cytokines that further amplify inflammation (innate immunity). Modified lipoproteins and heat shock proteins are identified as antigens by TRL4 in dendritic cells, leading to lymphocyte activation (adaptive immunity). Of note, a small portion of dendritic cells internalize lipoproteins and convert into foam cells [24]. The lymphocyte subsets located in atherosclerotic lesions are the T helper-1 (Th1), T helper-2 (Th2), and regulatory T cells (Threg). Th1 predominates in the lesions and exhibits pro-atherogenic properties by secreting IFNγ and TNFα. On the other hand, Th2 and Treg induce the production of atheroprotective molecules such as IL10, transforming growth factor β (TGFβ), IL4, and IL13 [25].

THE FORMATION OF ATHEROSCLEROTIC PLAQUE

The earliest histologically identified pro-atherosclerotic lesions are the fatty streaks formed by assembled foam cells in the intimal layer. They develop in artery segments with intimal thickening at a very early age. Fatty streaks can remain stable throughout life, progress into advanced atherosclerotic plaques, or sometimes even regress. Accumulated lipoproteins in the proteoglycan–collagen-enriched ECM form the initial progressive atherosclerotic lesion called *pathologic intimal thickening*. The infiltration of macrophages within the lipid pool (accumulated lipoproteins) and their subsequent apoptosis result in the development of the lipid-rich necrotic core, that combined with a collagen fibrous cap, constitutes the advanced lesion called *fibroatheroma* [26].

Apoptosis and Efferocytosis of Macrophages

Macrophage apoptosis in the intima plays an important role in the progression of atherosclerosis. Apoptotic cells are normally cleared by M2 macrophages and, to a lesser degree, by dendritic cells. The engulfment and ingestion of the apoptotic cells are known as efferocytosis, which serves the immediate and effective removal of apoptotic cells before loss of membrane integrity and release of inflammatory contents. Thus, efferocytosis counterbalances the harmful effects of apoptosis. This balance remains intact in early lesions but is disrupted in advanced plaques [27]. Triggering factors of apoptosis and defective efferocytosis are not fully identified. Oxidative stress and cholesterol accumulation seem to drive macrophages toward apoptosis through the stimulation of TRL2/4 signaling and prolonged ER stress. Defective efferocytosis may also result from oxidative stress or impaired function of receptor–ligand binding. Although the number of macrophages and consequently the inflammatory burden are reduced with apoptosis, in cells with defective efferocytosis the size of the necrotic core is increased by the necrotic cell residues [28]. Consequently, the imbalance between apoptosis and efferocytosis has detrimental effects on the progression of atherosclerotic lesions.

The Role of Vascular Smooth Muscle Cells

VSMCs also play a pivotal role in all stages of atherosclerotic disease. They normally reside in the media tunica and regulate vascular tone. Certain stimuli such as the subendothelium lipid accumulation, shear stress, and endothelial and macrophage-derived cytokines signal the switch of VSMCs from a quiescent "contractile" to a "synthetic" phenotype. The VSMCs' "switch" results in their migration from the media to the intima tunica, where they proliferate and produce proteoglycans and collagen. The enrichment of the ECM environment with proteoglycans contributes to LDL retention, whereas collagen deposition is implicated in intimal thickening and fibrous cap formation. In addition, VSMCs release a number of adhesion molecules and cytokines involved in monocyte recruitment and macrophage proliferation such as PDGF, TGFβ, and MCP1. Similar to macrophages, a small number of VSMCs also internalize modified LDL particles and turn into foam-like cells [29].

As mentioned in this chapter, the ECM plays a significant role in all atherosclerotic stages, beginning with the formation of intimal thickening and the retention of lipoproteins and extending to the development of the fibrous cap. ECM molecules are produced by VSMCs in response to PDGF and TGFβ stimuli. The degradation of ECM is performed by metalloproteinases (MMPs), which are endopeptidases secreted mainly by activated macrophages. ECM dissolution facilitates the transmigration and proliferation of cells during atherosclerotic progression. Furthermore, the arterial remodeling in response to atherosclerotic burden is sculpted by MMPs that fragment the internal elastic lamina [30,31]. The arterial architecture is modified in a way that plaque grows outwardly and thus does not result in lumen narrowing (positive or expansive remodeling). When plaque burden reaches a certain point, the wall cannot dilate any further and starts to expand into the lumen (constrictive remodeling), resulting in stenosis and flow obstruction [32].

Neovascularization

Formation of new vessels in the arterial wall, known as *neovascularization*, contributes to progression of atherosclerosis. Neovascularization is mainly triggered by the hypoxia attributed to intimal thickening and increased plaque size. As the diffuse distance between the surface and inner layer grows bigger, the oxygen supply in distal regions becomes less efficient. In addition, the activation of macrophages results in increased oxygen demands, further aggravating hypoxia and linking inflammation with neovascularization [33]. Under these conditions, the hypoxia inducible factor-1 (HIF1) triggers the production of vascular endothelial growth factor (VEGF) that signals ECs and VSMCs to develop new vessels [34]. This process is mostly associated with more advanced atherosclerotic lesions, but reported data confirm the existence of angiogenesis in early atheromatous lesions [35]. The formation of new vessels provides the atherosclerotic lesion with oxygen and nutrient supplies and thus promotes plaque growth, bypassing the restriction of diffuse distance. Furthermore, newly formed vessels are prone to rupture due to compromised structure integrity leading to intraplaque hemorrhage and plaque growth. It is quite clear that neovascularization and atherosclerosis are intertwined as they nurture one another [36].

CLINICAL IMPLICATIONS OF ATHEROSCLEROSIS

Atherosclerosis is a disease that initiates very early in life and progresses subclinically for many decades. Once the atherosclerotic plaque is formed, it can either remain asymptomatic throughout life or progress and cause symptoms of CAD.

The atherosclerotic plaque gradually expands, occluding the lumen. Symptoms mainly occur in stenosis greater than 60% under conditions of increased oxygen demand. In such cases, the patient presents with angina pectoris. Unfortunately, the first manifestation of CAD can also be an acute myocardial syndrome (ACS), including ST elevation myocardial infarction (STEMI), non-ST elevation myocardial infarction (NSTEMI), unstable angina (AU), or even sudden cardiac death. Lesions that cause critical stenosis of the lumen have a high probability to cause MI compared to noncritical coronary lesions, but as the latter are more frequently encountered in the vessel, they account for the majority of infarctions.

According to data from autopsy studies, the pathogenetic mechanism of ACS is a rupture or an erosion of the plaque with subsequent local thrombus formation that can totally or partially occlude the lumen flow, resulting in STEMI or NSTEMI/AU, respectively. Rupture is the underlying pathology in 75% of ACS, with a higher incidence in men and postmenopausal women, whereas erosion is more frequent among younger women and diabetic patients [37].

Atherosclerotic plaques are formed by a lipid-rich necrotic core and a fibrous cap that separates the core from the lumen and thus averts the contact of the thrombogenic core compounds with blood flow. Ruptured plaques are characterized by a weakened fibrous cap susceptible to mechanical hemodynamic forces. Collagen concentration is crucial in order for the fibrous cap to withstand those forces. Increased MMP activity in combination with decreased collagen synthesis by the VSMCs (increased apoptosis) account for the fibrous cap thinning. A large necrotic core is associated with plaque rupture due to the anatomical-mechanical changes it provokes and its enhanced inflammatory activity. Of note, the shoulder region of the plaque where rupture most frequently occurs is characterized by increased macrophage infiltration. Erosion is less studied and probably results from endothelium loss. The characteristics of vulnerable plaques (prone to rupture) offer valuable information regarding the pathogenetic mechanisms of ACS and set new therapeutic targets. According to current knowledge, a thin fibrous cap (thickness less than 65 μm), increased plaque burden (>40% of the size of the plaque), intense inflammatory activity and infiltration by macrophages, positive remodeling, a limited number of smooth muscle cells, increased neovascularization, and intraplaque hemorrhage characterize vulnerable plaques [38].

After rupture or erosion, the thrombogenic material of the necrotic core is exposed to the bloodstream. The size of the thrombus and the subsequent artery occlusion probably depend on the following variables, known as

Virchow's triad: (a) thrombogenicity of the exposed intima, (b) local flow, and (c) systematic predisposition to thrombosis. Irreversible myocardial cell damage occurs just 20 min from the onset of MI and is completed within 6 h. The infarct size is of particular clinical significance because myocardial necrosis greater than 25% or 40% usually leads to heart failure and cardiogenic shock, respectively. The extend of myocardial necrosis depends on which artery is occluded, the proximal or distal location of the culprit lesion, the time interval until blood flow restoration, and the presence of collateral circulation.

It is worth mentioning that not all plaque ruptures result in clinically symptomatic infarctions. Subclinical ruptures and thrombus formation boost VSMC proliferation, migration, and ECM production of ECM molecules, resulting in plaque size increase. This mechanism may explain the transition of a considerable percentage of thin-cap fibrous atheromas into thick-cap fibroatheromas during a 12-month follow-up [39].

GENETICS OF CORONARY ARTERY DISEASE

Despite the tremendous progress in primary and secondary prevention, CAD is unfortunately among the leading causes of death globally. Atherosclerosis, the underlying pathology of CAD, results from a combination of multiple pathophysiologic pathways, many of which remain unclear. The role of genetic factors in the occurrence of CAD has been the subject of intense research during recent decades. According to reported data, genetic factors can account for 40–60% of CAD predisposition. Results from a twin registry report a 44% as opposed to a 14% incidence of CAD between monozygotic and dizygotic twins [40]. Large-scale studies, such as the Framingham and Interheart Studies, point out that the family history of CAD is an independent risk factor for CAD [41,42]. Family history of CAD is incorporated in risk stratification scores such as the Reynolds risk score.

The complexity of atherosclerotic mechanisms is indicative of polygenic inheritance. In addition, the increased prevalence of CAD in the general population points to common genetic variants (>5% in the general population). Therefore, the studies for genetic variability in CAD should investigate a large population sample and include not only CAD patients but healthy subjects as well. However, the study design must examine a substantial number of genomic markers (ie, a polygenic pattern). Genome-wide association studies (GWAS) meet these specific study requirements, and in the last few years they have provided valuable insights into the genomics of CAD [43].

To date, 50 genes identified with GWAS have been associated with CAD. Although that is a significant number, each one contributes weakly to the calculation of cardiovascular risk. Therefore, it is quite discouraging that the total of 50 genes can account for less than 10% of the genetic variation associated with CAD. Fifty percent of the CAD-associated variants are encountered in over 50% of the population. It is noteworthy that only 15 genes out of the 50 are involved in the regulation of known cardiovascular risk factors, such as arterial hypertension and cholesterol and triglyceride levels. The remaining 35 genes promote atherosclerosis through as-yet-unknown mechanisms.

Interestingly, the majority of genes identified with GWAS are implicated in the development of early atherosclerosis. Genetic factors probably have relatively stronger effects on the early phases of atherosclerosis, where there is usually a marked absence of confounding factors such as arterial hypertension, diabetes mellitus, lack of exercise, and other comorbidities. Identification of genes associated with the risk of MI is still lacking. The only variants confirmed to predispose to MI – without, however, being associated with atherosclerosis – are the A and B blood type ones [43] (Table 2.1).

ABO Blood Groups

ABO blood group antigens are expressed mainly on the surface of red blood cells. According to studies, non-O blood type is associated with the occurrence of MI. The underlying mechanism is related not to atherosclerotic complications but to impaired hemostasis. In particular, blood groups A and B exhibit increased levels of von Willebrand factor (VWF) compared to O blood type (VWF levels gradually decline from the AB to B to A to O blood groups) [44]. Although the detailed mechanism linking blood type to VWF is not fully elucidated, a higher half-life of VWF is found in A and B blood groups, probably due to impaired proteolysis and clearance. The risk of MI in non-O blood types is approximately 20%. Supporting the association of non-O blood groups with increased thombogenicity, A and B blood types display a twofold risk of venous thrombosis [45].

Lipid-Associated Genes

Familiar hypercholesterolemia (FH) is an autosomal dominant disease characterized by high circulating LDL levels. In particular, in heterozygous carriers, the LDL levels are 2–3 times above the normal values and even higher in homozygotes (4–8 times) [46]. Carriers of gene mutations experience coronary events at a very early age (<30 years of age). The prevalence of FH is estimated at 1:500 and 1:1,000,000 persons for heterozygous and homozygous carriers, respectively. It is believed that 5% of heart attacks in people under the age of 60 can be attributed to heterozygous FH. Thus, early diagnosis and treatment with lipid-lowering interventions are of utmost importance. Genes implicated in the pathophysiology of FH by affecting LDL levels are LDL receptor (LDLR), PCSK9, apolipoprotein-B (apoB), and LDLR chaperone LDLR adaptor-related protein 1 (LDLRAP1). However, a considerable percentage (30–50%) of phenotypically diagnosed FH is not related to any of the identified genes, pointing to a polygenic pattern [47,48].

TABLE 2.1 Representative Genes of the Association With Coronary Artery Disease

	Chromosomal Location	Protein	Mechanisms of Action
Genes Related to LDL Cholesterol			
LDLR	19p13.2	Cell surface receptor of LDL	Internalization, transfer to lysosomes, and degradation to free cholesterol
PCSK9	1p32.3	Proprotein convertase	Regulates LDLR cleavage
LPA	6q25.3	Lipoprotein A	Affects fibrinolysis Unknown mechanisms for atherosclerotic implications
APOB	2p24.1	Structural apolipoprotein of LDL	Facilitates transportation of LDL; binds to LDL receptor
SORT1	1p13.3	Sortilin-1, multiligand receptor protein on cell membrane and intracellular compartments	Golgi–endosome sorting; VLDL production rates; LDL uptake; VSMC apoptosis; LDL uptake – conflicting data from animal studies
APOE	19q13.32	Structural component of triglyceride-rich lipoproteins	Transportation and delivery of lipids to various tissues and cells
ABCG5/ABCG8	2p21	ATP-binding cassette (ABC) transporter	Export of free sterols from hepatocytes and enterocytes to the lumen
Genes Related to Triglycerides			
APOA5	11q23.3	Apolipoprotein A5	Regulates triglyceride catabolism, secretion of VLDL, and lipoproteins' cell uptake
Genes Related to Myocardial Infarction			
ABO	9q34.2	ABO blood group antigens	Impaired hemostasis; increased levels of von Willebrand factor
Genes With Unknown Mechanism(s)			
CDKN2A/CDKN2B	9p21.3	Antisense noncoding RNA in the *INK4* locus (ANRIL)	Cell cycle and proliferation; unknown mechanisms

VLDL, Very low density lipoprotein; *LDL*, low density lipoprotein; *LDLR*, LDL receptor; *VSCM*, vascular smooth muscle cell.

Identification of the LDLR gene is considered a hallmark in the genetic field of CAD. It was some decades ago that the observation of an early and high occurrence of coronary events in certain family groups was combined with very high levels of plasma LDL. Further research revealed the critical role of the LDLR in lipid metabolism. Specifically, the LDLR molecule is a cell surface receptor for LDL molecules that mediates their internalization and transfer to the lysosomes in order to be degraded to free cholesterol. Increased intracellular cholesterol concentration signals the suppression of LDLR and cholesterol biosynthesis in order to maintain lipid homeostasis [48]. Loss-of-action mutations of LDLR genes lead to elevated plasma lipid levels (phenotype of FH) and are inherited as an autosomal dominant trait. Mutations of the LDLR gene are identified in 85–90% of cases of FH [49]. The discovery of LDLR mutation and the deciphering of lipid metabolism have led to the revolutionary development of statin therapy.

The identification of the LDLR loss-of-action mutation provided valuable insights into cholesterol metabolism and atherosclerosis treatment. More recently, a new protein involved in LDLR degradation came to light. PCSK9 is a proprotein convertase, mainly expressed in the liver as well as in the gastrointestinal and nervous systems, that regulates LDLR cleavage after binding to its epidermal growth factor domain. The PCSK9 gene is inherited as an autosomal dominant trait. Increased levels of circulating LDL and a high occurrence rate of coronary events are observed in subjects with a gain-of-action mutation of the PCSK9 gene. However, low lipid levels and a marked reduction (88%) in risk for CAD are seen in people carrying the loss-of-action mutation.

The role of lipoprotein-A (LPA) in atherosclerosis and cardiovascular risk has been under investigation for a long period of time. LPA consists of an LDL like-particle and an apolipoprotein-A (apoA) molecule that is bound to the apoB particle. LPA shows atherogenic action attributed to (a) the similarity of the molecule with plasminogen, (b) inactivating the tissue factor inhibitor pathway, and (c) causing endothelial dysfunction and exacerbation of inflammatory processes in atheromatous plaque. The apoA gene is the principal regulator of LPA concentration and accounts for the heterogeneity in plasma levels between various population groups. High levels of LPA are associated with increased cardiovascular risk [50]. apoB is a structural apolipoprotein of LDL particles that facilitates their transportation to the peripheral tissues. In addition, apoB binds to LDLR, ensuing the internalization of LDL into the cells. A mutation of the apoB gene, inherited in an autosomal dominant trait, results in a defective interaction with LDLR and high levels of circulating LDL.

Mutations of the liver-specific LDLRAP1 are associated with impaired LDL internalization, although the LDLR displays normal function. It is also known as *autosomal recessive hypercholesterolemia* and presents with an intermediate phenotype between the ones of classical heterozygous and homozygous FH [51].

Several additional genes that affect lipid levels have been identified in GWAS, such as SORT1, APOE, APO5, ABCG5, and ABCG8. Sortilin-1 is a multi–ligand receptor protein encoded by the SORT1 gene. It is expressed on the cell membrane (10%) and in intracellular compartments (90%) participating in Golgi–endosome sorting, VLDL production rates and LDL uptake. Recent data implicate SORT1 in VSMC apoptosis as well as in LDL uptake by macrophages in atherosclerotic lesion [52,53]. The interaction mechanism of SORT1 and LDL levels is still not fully understood, as animal studies present conflicting reports [54]. Apolipoprotein E (apoE) is a structural component of triglyceride-rich lipoproteins and is encoded by the polymorphic apoE gene. The three isoforms E2, E3, and E4 of the apoE gene display differences in frequency and functionality. The most frequent isoform is E3 (approximately 60%), and it is considered the normal one; E4 is associated with increased serum cholesterol levels and a frequency of 30%, whereas E2 is associated with decreased serum levels and a 10% frequency [55]. Mutations in the ABCG5 and ABCG8 genes are responsible for the autosomal recessively inherited disease sitosterolemia. Increased LDL levels are attributed to increased export of free sterols from hepatocytes and enterocytes to the lumen [56]. Gene mutations (apoA5) that increase triglyceride levels also modify cardiovascular risk [57].

Other Atherosclerosis-Related Genes

As mentioned, a total of 50 genetic variants have been associated with CAD, many of which modify cardiovascular risk and atherosclerosis progression through unknown mechanisms (Table 2.1). The most established genetic factor of CAD is the 9p21 locus. In 2007, GWAS reported the 9p21 locus as the first chromosome region related to CAD genomics [58]. Since then, many studies reciprocated the former results validating the association between 9p21 loci and CAD, independently of known risk factors. The mechanism connecting the 9p21 loci with cardiovascular risk is not yet clarified [59,60]. The 9p21 region does not contain protein-coding genes, but instead it harbors a long intergenic noncoding RNA called ANRIL (antisense noncoding RNA in the *INK4* locus). The expression of three nearby genes CDKN2A, CDKN2B, and MTAP that are involved in cell cycle and proliferation is induced by ANRIL. According to a recent meta-analysis, 9p21 loci are strongly associated with the first incidence of CAD but lack association with subsequent coronary events [61].

The Role of microRNAs

MicroRNAs (miRNAs) are small single-stranded noncoding RNAs that regulate gene expression in humans by inhibiting protein translation. Recent data highlight the role of miRNAs in almost all stages of the atherosclerotic process [62].

The interaction between endothelial cells and miRNAs is mostly mediated by hemodynamic stimuli [63]. Specifically, laminar flow increases miR-10a and mir-181b levels that exert atheroprotective properties by inhibiting NFκB activation [64,65]. On the contrary, in low-shear-stress sites, the upregulated mir-92a, mir-34, mir-21, and mir-712 promote endothelial inflammation by inhibiting PPAP-alpha messenger RNA (mRNA) and activating the NFκB pathway [66–68]. Levels of the atheroprotective mir-19a and mir-126 are downregulated by low shear stress and hyperlipidemia [69,70].

MiRNAs are implicated also in the activation of macrophages and the inflammatory cascade. M1 macrophages exhibit increased levels of mir-155 and mir-342-5p that in turn promote inflammatory gene expression [71,72]. On the contrary, the expression of mir-24 exhibits antiinflammatory functions and limits atherosclerosis [73]. In addition, miRNAs are involved in the atheroprotective mechanism of reverse cholesterol transport. In particular, mir-19b, mir-33, and mir-144-3p increase the accumulation of cholesterol in macrophages by targeting the ABCG1 and ABCA1 transporters [74,75].

The regulation of several key functions by miRNAs provides the foundation for new therapeutic targets and may lead to the development of new biomarkers for cardiovascular risk stratification.

ATHEROSCLEROSIS: THERAPEUTIC OPTIONS

Large-scale studies validate the association between atherosclerosis and LDL, underscoring the fact that increased circulated levels of LDL are associated with elevated cardiovascular risk. Subsequently to dietary and lifestyle changes, the administration of therapeutic agents that reduce LDL levels is the first-line treatment of atherosclerosis.

Statins are considered the standard lipid-lowering therapy, and their use in clinical practice drastically altered the therapeutic approach toward atherosclerosis and CAD. The reduction in LDL levels is mediated through the inhibition of a rate-limiting enzyme in cholesterol synthesis called hydroxymethylglutaryl–coenzyme A reductase. Statins also exert pleiotropic antiinflammatory effects (plaque stabilization, improved endothelial function, and inhibition of platelet activation) independently of their lipid-lowering properties that contribute to the improvement of cardiovascular risk. Numerous studies have demonstrated the benefits of statin therapy in primary and secondary prevention of CAD [76–78]. According to a recent meta-analysis, people who may benefit more from statin administration are those with low risk in contrast to primary beliefs. This observation probably reflects a better efficacy of statins in early lesions [79].

Despite the achieved reduction in LDL levels, many other lipid-lowering agents failed to decrease the occurrence of cardiovascular events [80,81]. The so-far-irreplaceable benefits of statin administration may be attributed either to its pleiotropic effects or to the specific mechanism by which LDL is reduced. Lomitapide, mipomersen, and the PCSK9 inhibitors are novel antilipidemic agents currently under clinical investigation. The PCSK9 inhibitors (monoclonal antibodies alirocumab, bococizumab, and evolocumab) emerge as very promising novel agents with no evident major side effects. The reduction in LDL levels is so profound that the main concern of researchers is to define the safest lower LDL value [82]. Mipomersen and lomitapide are two agents, each with a unique mechanism of action, that have been approved recently by the US Food and Drug Administration for the treatment of FH. Mipomersen is a single-stranded, antisense oligonucleotide that binds to a mRNA that encodes apolipoprotein B-100. Lomitapide inhibits the microsomal triglyceride transfer protein (MTP), resulting in the reduction of chylomicrons and VLDL [83].

Inflammation plays a critical role in atherosclerosis progression. Therefore, agents such as methotrexate, colchicine, and allopurinol that are known for their antiinflammatory effects are tested for atheroprotective properties. Methotrexate is a folic acid antagonist administered as a chemotherapeutic agent for cancer and in low doses as an antiinflammatory agent in autoimmune disorders. A low Methotrexate dose reduced cardiovascular risk in a population of rheumatoid arthritis patients [84]. Colchicine is a pharmaceutical agent used to treat arthritis and gout. The improvement of cardiovascular prognosis in patients with CAD receiving colchicine could be attributed to the inhibition of NLRP3 [85,86]. The increased cardiovascular risk in patients with autoimmune diseases may be explained by the chronic inflammatory process. The administration of antibodies inhibiting TNF (eg, infliximab and adalimumab) was associated with a decrease in coronary events in patients with rheumatoid arthritis [87]. Canakinumab is an IL1b-neutralizing monoclonal antibody that is currently under evaluation for its antiinflammatory effects in a stable CAD population study (CANTOS trial) [88]. Other agents targeting inflammatory components that are currently under investigation are losmapimod (p38MARK inhibitor), MLN1202 (monoclonal antibody against CCR2), and briakinumab (monoclonal antibody against IL12) [89]. Interestingly, vaccination with oxidized LDL particles has proven atheroprotective in animal studies [90].

The deciphering of basic atherosclerotic mechanisms in combination with cardiovascular genetics provided a significant number of novel therapeutic targets. The use of miRNAs as pharmacological targets in atherosclerosis is very promising. MiRNA mimics are still in development, whereas anti-miRNAs (antagomirs) are used to inhibit miRNAs [91,92]. Genetic screening is not included in risk stratification scores as it does not increase the prognostic value when added to traditional risk factors. However, according to recently published data, a genetic risk score of 27 variants identified individuals at increased risk for both first-incident and recurrent coronary heart disease events. In addition, the group of people at high genetic risk was the one to benefit more from statin treatment [76]. Finally, certain genetic polymorphisms can modify the response to therapeutic agents administered in patients with CAD. For example, the P450 CYP2C19*2 polymorphism is associated with clopidogrel resistance and recurrence of adverse cardiovascular events [93] whereas the SLCO1B1*5 allele is associated with statin myopathy [94], posing the question of genetic screening.

REFERENCES

[1] Laslett LJ. The worldwide environment of cardiovascular disease: prevalence, diagnosis, therapy, and policy issues: a report from the American College of Cardiology. J Am Coll Cardiol 2012;60:S1–49.

[2] Strong JP. Prevalence and extent of atherosclerosis in adolescents and young adults: implications for prevention from the pathobiological determinants of atherosclerosis in youth study. JAMA 1999;281:727–35.

[3] Mahmood SS. The Framingham heart study and the epidemiology of cardiovascular disease: a historical perspective. Lancet 2014;383:999–1008.

[4] Greenland P. Major risk factors as antecedents of fatal and nonfatal coronary heart disease events. JAMA 2003;290:891–7.

[5] Deanfield JE. Endothelial function and dysfunction: testing and clinical relevance. Circulation 2007;115:1285–95.

[6] Majesky MW. The adventitia: a dynamic interface containing resident progenitor cells. Arterioscler Thromb Vasc Biol 2011;31:1530–9.

[7] Chatzizisis YS. Role of endothelial shear stress in the natural history of coronary atherosclerosis and vascular remodeling: molecular, cellular, and vascular behavior. J Am Coll Cardiol 2007;49:2379–93.

[8] Vanhoutte PM. Endothelial dysfunction: the first step toward coronary arteriosclerosis. Circ J 2009;73:595–601.

[9] Tsimikas S. Oxidative modification of lipoproteins: mechanisms, role in inflammation and potential clinical applications in cardiovascular disease. Curr Pharm Des 2011;17:27–37.

[10] Williams KJ. The response-to-retention hypothesis of early atherogenesis. Arterioscler Thromb Vasc Biol 1995;15:551–61.

[11] Moore KJ. Macrophages in atherosclerosis: a dynamic balance. Nat Rev Immunol 2013;13:709–21.

[12] Chinetti-Gbaguidi G. Macrophage subsets in atherosclerosis. Nat Rev Cardiol 2015;12:10–7.

[13] Mosser DM. The many faces of macrophage activation. J Leukoc Biol 2003;73:209–12.

[14] Zizzo G. Efficient clearance of early apoptotic cells by human macrophages requires M2c polarization and MerTK induction. J Immunol 2012;189:3508–20.

[15] Porcheray F. Macrophage activation switching: an asset for the resolution of inflammation. Clin Exp Immunol 2005;142:481–9.

[16] Kunjathoor VV. Scavenger receptors class A-I/II and CD36 are the principal receptors responsible for the uptake of modified low density lipoprotein leading to lipid loading in macrophages. J Biol Chem 2002;277:49982–8.

[17] Maxfield FR. Role of cholesterol and lipid organization in disease. Nature 2005;438:612–21.

[18] Yu XH. Foam cells in atherosclerosis. Clin Chim Acta 2013;424:245–52.

[19] Weber C. Atherosclerosis: current pathogenesis and therapeutic options. Nat Med 2011; 17:1410–22.

[20] Spann NJ. Regulated accumulation of desmosterol integrates macrophage lipid metabolism and inflammatory responses. Cell 2012;151:138–52.

[21] Murphy AJ. Proliferating macrophages populate established atherosclerotic lesions. Circ Res 2014;114:236–8.

[22] van Gils JM. The neuroimmune guidance cue netrin-1 promotes atherosclerosis by inhibiting the emigration of macrophages from plaques. Nat Immunol 2012;13:136–43.
[23] Wanschel A. Neuroimmune guidance cue semaphorin 3E is expressed in atherosclerotic plaques and regulates macrophage retention. Arterioscler Thromb Vasc Biol 2013;33:886–93.
[24] Bobryshev YV. Dendritic cells and their role in atherogenesis. Lab Investig J Tech Methods Pathol 2010;90:970–84.
[25] Tse K. T cells in atherosclerosis. Int immunol 2013;25:615–22.
[26] Virmani R. Vulnerable plaque: the pathology of unstable coronary lesions. J Interv Cardiol 2002;15:439–46.
[27] Moore KJ. Macrophages in the pathogenesis of atherosclerosis. Cell 2011;145:341–55.
[28] Tabas I. Macrophage death and defective inflammation resolution in atherosclerosis. Nat Rev Immunol 2010;10:36–46.
[29] Johnson JL. Emerging regulators of vascular smooth muscle cell function in the development and progression of atherosclerosis. Cardiovasc Res 2014;103:452–60.
[30] Siasos G. Inflammatory mechanisms in atherosclerosis: the impact of matrix metalloproteinases. Curr Top Med Chem 2012;12:1132–48.
[31] Koskinas KC. Natural history of experimental coronary atherosclerosis and vascular remodeling in relation to endothelial shear stress: a serial, in vivo intravascular ultrasound study. Circulation 2010;121:2092–101.
[32] Toutouzas K. Matrix metalloproteinases and vulnerable atheromatous plaque. Curr Top Med Chem 2012;12:1166–80.
[33] Nakashima Y. Distributions of diffuse intimal thickening in human arteries: preferential expression in atherosclerosis-prone arteries from an early age. Virchows Arch 2002;441:279–88.
[34] Kuwahara F. Hypoxia-inducible factor-1alpha/vascular endothelial growth factor pathway for adventitial vasa vasorum formation in hypertensive rat aorta. Hypertension 2002;39:46–50.
[35] Ho-Tin-Noe B. Early atheroma-derived agonists of peroxisome proliferator-activated receptor-gamma trigger intramedial angiogenesis in a smooth muscle cell-dependent manner. Circ Res 2011;109:1003–14.
[36] Xu J. Vasa vasorum in atherosclerosis and clinical significance. Int J Mol Sci 2015;16:11574–608.
[37] Libby P. Mechanisms of acute coronary syndromes and their implications for therapy. N Engl J Med 2013;368:2004–13.
[38] Stone GW. A prospective natural-history study of coronary atherosclerosis. N Engl J Med 2011;364:226–35.
[39] Falk E. Update on acute coronary syndromes: the pathologists' view. Eur Heart J 2013;34:719–28.
[40] Allen G. Measures of twin concordance. Acta genetica et statistica medica 1967;17:475–81.
[41] Schildkraut JM. Coronary risk associated with age and sex of parental heart disease in the Framingham study. Am J Cardiol 1989;64:555–9.
[42] Yusuf S. Effect of potentially modifiable risk factors associated with myocardial infarction in 52 countries (the INTERHEART study): case-control study. Lancet 2004;364:937–52.
[43] Roberts R. Genetics of coronary artery disease. Circ Res 2014;114:1890–903.
[44] Jenkins PV. ABO blood group determines plasma von Willebrand factor levels: a biologic function after all? Transfusion 2006;46:1836–44.
[45] Franchini M. ABO blood group and thrombotic vascular disease. Thromb Haemost 2014;112:1103–9.
[46] Maxfield FR. Cholesterol, the central lipid of mammalian cells. Curr Opin Cell Biol 2010;22:422–9.
[47] Ahmad Z. Low prevalence of mutations in known loci for autosomal dominant hypercholesterolemia in a multiethnic patient cohort. Circ Cardiovasc Genet 2012;5:666–75.

[48] Sniderman AD. The severe hypercholesterolemia phenotype: clinical diagnosis, management, and emerging therapies. J Am Coll Cardiol 2014;63:1935–47.
[49] Haase A. Identification of people with heterozygous familial hypercholesterolemia. Curr Opin Lipidol 2012;23:282–9.
[50] Durrington PN. Lipoprotein (a): gene genie. Curr Opin Lipidol 2014;25:289–96.
[51] Soutar AK. Mechanisms of disease: genetic causes of familial hypercholesterolemia. Nat Clin Pract Cardiovasc Med 2007;4:214–25.
[52] Campagnolo L. Sortilin expression is essential for pro-nerve growth factor-induced apoptosis of rat vascular smooth muscle cells. PLoS One 2014;9:e84969.
[53] Patel KM. Macrophage sortilin promotes LDL uptake, foam cell formation, and atherosclerosis. Circ Res 2015;116:789–96.
[54] Strong A. Sortilin and lipoprotein metabolism: making sense out of complexity. Curr Opin Lipidol 2014;25:350–7.
[55] Mahley RW. Apolipoprotein E: far more than a lipid transport protein. Annu Rev Genom Hum Genet 2000;1:507–37.
[56] Escola-Gil JC. Sitosterolemia: diagnosis, investigation, and management. Curr Atheroscler Rep 2014;16:424.
[57] Sharma V. Influence of apolipoprotein A-V on the metabolic fate of triacylglycerol. Curr Opin Lipidol 2013;24:153–9.
[58] McPherson R. A common allele on chromosome 9 associated with coronary heart disease. Science 2007;316:1488–91.
[59] Tousoulis D. Candidate gene polymorphisms and the 9p21 locus in acute coronary syndromes. Trends Mol Med 2008;14:441–9.
[60] Munir MS. The association of 9p21-3 locus with coronary atherosclerosis: a systematic review and meta-analysis. BMC Med Genet 2014;15:66.
[61] Patel RS. Genetic variants at chromosome 9p21 and risk of first versus subsequent coronary heart disease events: a systematic review and meta-analysis. J Am Coll Cardiol 2014;63:2234–45.
[62] Andreou I. miRNAs in atherosclerotic plaque initiation, progression, and rupture. Trends Mol Med 2015;21:307–18.
[63] Tousoulis D. Novel biomarkers assessing endothelial dysfunction: role of microRNAs. Curr Top Med Chem 2013;13:1518–26.
[64] Fang Y. MicroRNA-10a regulation of proinflammatory phenotype in athero-susceptible endothelium in vivo and in vitro. Proc Natl Acad Sci U S A 2010;107:13450–5.
[65] Sun X. MicroRNA-181b regulates NF-kappaB-mediated vascular inflammation. J Clin Investig 2012;122:1973–90.
[66] Zhou J. MicroRNA-21 targets peroxisome proliferators-activated receptor-alpha in an autoregulatory loop to modulate flow-induced endothelial inflammation. Proc Natl Acad Sci U S A 2011;108:10355–60.
[67] Loyer X. Inhibition of microRNA-92a prevents endothelial dysfunction and atherosclerosis in mice. Circ Res 2014;114:434–43.
[68] Son DJ. The atypical mechanosensitive microRNA-712 derived from pre-ribosomal RNA induces endothelial inflammation and atherosclerosis. Nat Commun 2013;4:3000.
[69] Qin X. MicroRNA-19a mediates the suppressive effect of laminar flow on cyclin D1 expression in human umbilical vein endothelial cells. Proc Natl Acad Sci U S A 2010;107:3240–4.
[70] Schober A. MicroRNA-126-5p promotes endothelial proliferation and limits atherosclerosis by suppressing Dlk1. Nat Med 2014;20:368–76.
[71] Tian FJ. Elevated microRNA-155 promotes foam cell formation by targeting HBP1 in atherogenesis. Cardiovasc Res 2014;103:100–10.

[72] Wei Y. The microRNA-342-5p fosters inflammatory macrophage activation through an Akt1- and microRNA-155-dependent pathway during atherosclerosis. Circulation 2013;127:1609–19.
[73] Di Gregoli K. MicroRNA-24 regulates macrophage behavior and retards atherosclerosis. Arterioscler Thromb Vasc Biol 2014;34:1990–2000.
[74] Davalos A. From evolution to revolution: miRNAs as pharmacological targets for modulating cholesterol efflux and reverse cholesterol transport. Pharmacol Res 2013;75:60–72.
[75] Schober A. MicroRNA-mediated mechanisms of the cellular stress response in atherosclerosis. Nat Rev Cardiol 2015;12:361–74.
[76] Mega JL. Genetic risk, coronary heart disease events, and the clinical benefit of statin therapy: an analysis of primary and secondary prevention trials. Lancet 2015;385:2264–71.
[77] Ridker PM. Rosuvastatin to prevent vascular events in men and women with elevated C-reactive protein. N Engl J Med 2008;359:2195–207.
[78] Tousoulis D. Innate and adaptive inflammation as a therapeutic target in vascular disease: the emerging role of statins. J Am Coll Cardiol 2014;63:2491–502.
[79] Mihaylova B. The effects of lowering LDL cholesterol with statin therapy in people at low risk of vascular disease: meta-analysis of individual data from 27 randomised trials. Lancet 2012;380:581–90.
[80] Barter PJ. Effects of Torcetrapib in patients at high risk for coronary events. N Engl J Med 2007;357:2109–22.
[81] Schwartz GG. Effects of Dalcetrapib in patients with a recent acute coronary syndrome. N Engl J Med 2012;367:2089–99.
[82] Seidah NG. The biology and therapeutic targeting of the proprotein convertases. Nat Rev Drug Discov 2012;11:367–83.
[83] Ridker PM. LDL cholesterol: controversies and future therapeutic directions. Lancet 2014;384:607–17.
[84] Westlake SL. The effect of methotrexate on cardiovascular disease in patients with rheumatoid arthritis: a systematic literature review. Rheumatology (Oxford) 2010;49:295–307.
[85] Martinon F. Gout-associated uric acid crystals activate the NALP3 inflammasome. Nature 2006;440:237–41.
[86] Nidorf SM. Low-dose colchicine for secondary prevention of cardiovascular disease. J Am Coll Cardiol 2013;61:404–10.
[87] Dixon WG. Reduction in the incidence of myocardial infarction in patients with rheumatoid arthritis who respond to anti-tumor necrosis factor alpha therapy: results from the British Society for Rheumatology Biologics Register. Arthritis Rheum 2007;56:2905–12.
[88] Ridker PM. Effects of interleukin-1beta inhibition with Canakinumab on hemoglobin A1c, lipids, C-reactive protein, interleukin-6, and fibrinogen: a phase IIb randomized, placebo-controlled trial. Circulation 2012;126:2739–48.
[89] Back M. Anti-inflammatory therapies for atherosclerosis. Nat Rev Cardiol 2015;12:199–211.
[90] Shah PK. Vaccine for atherosclerosis. J Am Coll Cardiol 2014;64:2779–91.
[91] Condorelli G. MicroRNAs in cardiovascular diseases: current knowledge and the road ahead. J Am Coll Cardiol 2014;63:2177–87.
[92] Siasos G. MicroRNAs in cardiovascular therapeutics. Curr Top Med Chem 2013;13:1605–18.
[93] Oikonomou E. Genetic testing and antiplatelet treatment: still way to go? Int J Cardiol 2015;187:63–5.
[94] Mangravite LM. A statin-dependent QTL for GATM expression is associated with statin-induced myopathy. Nature 2013;502:377–80.

Chapter 3

The Role of Genetics in Acute Coronary Syndrome

J. Chandrasekhar, R. Mehran
Icahn School of Medicine at Mount Sinai, New York, NY, United States

INTRODUCTION

Acute coronary syndrome (ACS) is a significant health-care problem and despite medical advances accounts for considerable morbidity and mortality each year. In the United States >780,000 individuals each year are diagnosed with ACS, at a median age of 68 years and a male-to-female ratio of approximately 3:2 [1]. The cornerstone of ACS treatment is pharmacotherapy with antithrombotic agents. Dual antiplatelet therapy (DAPT) with aspirin and a $P2Y_{12}$ receptor inhibitor is recommended for a period of 12 months irrespective of invasive revascularization [1]. Three contemporary oral $P2Y_{12}$ receptor antagonists, namely clopidogrel, prasugrel, and ticagrelor are available for use with aspirin in patients with ACS. In the Platelet Inhibition and Patient Outcomes (PLATO) study, ticagrelor resulted in a 16% relative risk reduction (RRR) in major adverse cardiovascular events (MACE) and a 24% RRR in mortality compared with clopidogrel in ACS patients [2]. Similarly, prasugrel was superior to clopidogrel in ACS patients undergoing percutaneous coronary intervention (PCI) in the TRITON TIMI 38 study, with a significant reduction in MACE (19% RRR) and stent thrombosis (54% RRR) [3].

Consistent with advances in stent design, adverse ischemic outcomes are now dramatically low in both low- and high-risk patients undergoing PCI [4]. Conversely, with the prescription of potent pharmacotherapy, bleeding outcomes have come under scrutiny. A desirable antiplatelet regimen is one that not only reduces MACE but also bleeding outcomes for a net clinical benefit. Thus patient selection and a personalized management approach for the optimal balance of efficacy and safety have become critically important. It is in this context that we discuss the pharmacokinetic and pharmacodynamic differences exhibited by antithrombotic agents along with the influence of interindividual and racial genetic variations. This chapter discusses the current genetic evidence principally relevant to antiplatelet therapy, as well as the potential role of genotyping in prescription of antiplatelet therapies in ACS.

Cardiovascular Diseases.

CLOPIDOGREL

DAPT with aspirin and clopidogrel first demonstrated significant ischemic benefit in the randomized CURE trial compared with aspirin monotherapy, in ACS and including patients undergoing PCI [5,6]. Subsequently use of clopidogrel became standard practice in the management of a significant proportion of patients with coronary artery disease. Clopidogrel is a second-generation thienopyridine and produces adenosine diphosphate (ADP)-mediated antithrombotic effect by irreversibly binding to the $P2Y_{12}$ platelet receptor. It is a prodrug and requires conversion to the active thiol metabolite. This is subject to significant interindividual variation, and the resultant differences in pharmacokinetics including intestinal absorption and hepatic oxidation have an ultimate influence on pharmacodynamic as well as clinical outcomes.

The role of genetic variation led the US Food and Drug administration to issue a black box warning in 2010 for consideration in patients identified as poor or nonresponders to clopidogrel as these patients are potentially at a heightened risk of ischemic events (http://www.fda.gov/drugs/drugsafety/postmarketdrugsafetyinformationforpatientsandproviders/ucm203888.htm). Since then extensive data has become available on this subject with consensus on some genetic polymorphisms and conflicting results on others. This section discusses clopidogrel metabolism and the evidence on specific genetic polymorphisms.

Metabolism and Pharmacokinetics

Absorption and Role of P-glycoprotein

Intestinal absorption of clopidogrel is affected by the transporter P-glycoprotein encoded by the ABCB1 or multidrug resistance 1 (MDR1) gene located on the long arm of chromosome 7 at position 21 (7q21) [7]. P-glycoprotein is an ATP dependent efflux pump variably expressed in the population and located at the intestinal surfaces, as well as hepatic, renal surfaces and the blood–brain barrier. At a cellular level, it is expressed on the plasma membrane, Golgi membrane, and intracellular canaliculus, and its main function is to prevent absorption by promoting efflux. It is linked to the CYP3A4 isoform of cytochrome P450 enzymes with a propensity to be induced or inhibited. Primarily protective from xenobiotic agents, it is associated with multidrug resistance in cancer cells. Modulation of the drug-binding sites of P-glycoprotein thus results in unique transport patterns for each drug. In the case of clopidogrel, induction of P-glycoprotein reduces absorption.

Hepatic Carboxylation and Oxidation

Up to 85% of clopidogrel reaching the liver is hydrolyzed by esterases in a shunting step to an inactive carboxylic acid derivative while the remaining 15% is available for activation (Fig. 3.1). Conversion to the active metabolite

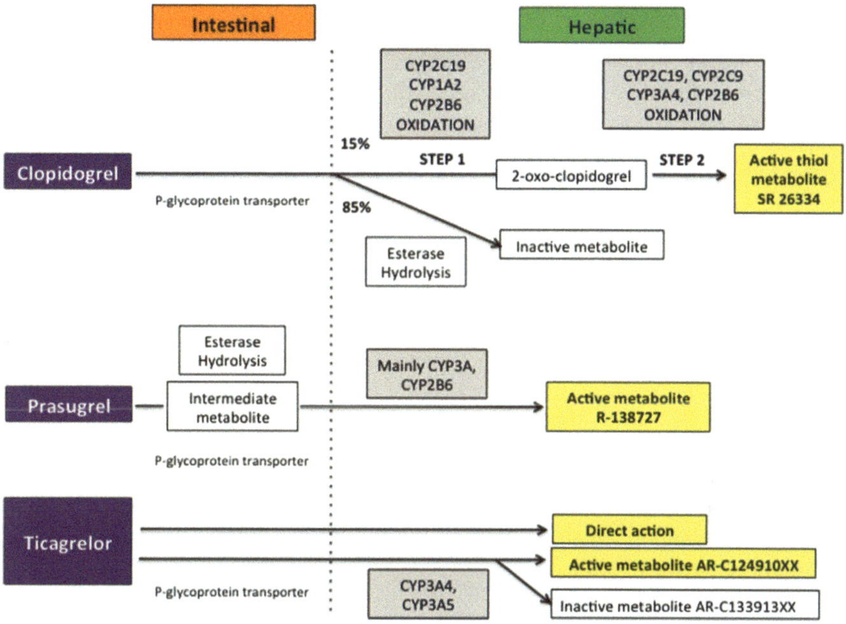

FIGURE 3.1 Metabolic pathways for clopidogrel, prasugrel, and ticagrelor.

is a two-step oxidation process by cytochrome P450 enzymes. The first step involves oxidation to 2-oxo-clopidogrel predominantly by CYP1A2, CYP2B6, and CYP2C19. Further oxidation to the active thiol metabolite (SR 26334) is brought about by CYP3A4, CYP2B6, CYP2C9, and CYPC19 [8]. Thus CYP2C19 plays a crucial role in both activation steps, contributing to 45% in the generation of 2-oxo-clopidogrel in the first step and 29% in the generation of thiol active metabolite in the second step. The active clopidogrel metabolite irreversibly inhibits the $P2Y_{12}$ platelet receptor for the lifespan of the platelet and reduces ADP-mediated platelet aggregation by ~50%. The time to peak plasma levels after a loading dose can be several hours [9]. Reduced availability of the active clopidogrel metabolite results in lower inhibition of platelet aggregation and the phenomenon of "clopidogrel resistance."

Effect of Genotypic Polymorphisms

Several genetic variations alter the pharmacokinetics and pharmacodynamics of clopidogrel as discussed later. However, the most significant variation appears to be the single nucleotide polymorphism (SNP) in chromosome 10 resulting in the CYP2C19*2 allele. Table 3.1 indicates select genetic polymorphisms related to antiplatelet therapies.

TABLE 3.1 Select Genetic Polymorphisms Related to Antiplatelet Therapy

Gene/Allele/SNP	Reference SNP ID Number
P-glycoprotein Transporter	
ABCB1/Multidrug resistance 1 (MDR1), C3435T	rs1045642
Cytochrome P450 Enzymes	
CYP2C19*1	Wild type
CYP2C19*2, G681A	rs4244285
CYP2C19*3	*rs4986893*
CYP2C19*4	rs28399504
CYP2C19*5	*rs56337013*
CYP2C19*6	rs72552267
CYP2C19*7	rs72558186
CYP2C19*8	rs41291556
CYP2C19*9	rs17884712
CYP2C19*10	rs6413438
CYP2C19*17	rs12248560
Paraoxonase 1, Q192R	rs662
Platelet Surface Receptors	
$P2Y_{12}$ receptor gene, T744C	rs2046934
Platelet endothelial aggregation receptor-1 (PEAR1)	rs12041331
Protease-activated receptor 1 (PAR1)	rs168753
Protease-activated receptor 4 (PAR4), F2RL3	rs773902

SNP, Single nucleotide polymorphism.

ABCB1

The commonest ABCB1 polymorphism is a C>T change at position 3435 (ABCB1 3435C>T). Taubert et al. identified that patients with induced P-glycoprotein had lower levels of area under the concentration–time curve of the active clopidogrel metabolite [7]. In a study of 401 Chinese ACS patients both the incidence of ABCB1 3435TT gene as well as CYP2C19*2 allele were associated with reduced level of active and inactive clopidogrel metabolites and lower platelet inhibition on light transmission aggregometry [10].

In healthy volunteers treated with a 300 or 600 mg clopidogrel loading dose, the presence of ABCB1 3435C>T was associated with a reduced pharmacodynamic suppression by 7.2% in TT homozygotes compared with CT/CC variants [11].

In the TRITON TIMI 38 study, 2932 patients were genotyped for ABCB1 polymorphisms—the incidence of TT homozygosity was 27%, CT heterozygosity was 50%, and CC homozygosity was 23% [11]. The incidence of the primary MACE endpoint (death, MI, stroke, or unplanned revascularization) until 15 months was 72% higher in the TT homozygotes compared with the CC homozygotes or CT heterozygotes on clopidogrel (Kaplan–Meier event rates 12.9% vs 7.8%, HR 1.72, 95% CI 1.22–2.44, $p=.002$). While the hazard ratios (HRs) were not significant for death, stent thrombosis or nonfatal stroke, the HR for nonfatal myocardial infarction (MI) was 1.82 (12.0% vs 6.8%, 95% CI 1.26–2.62, $p=.0013$). This difference was noted as early as 30 days and persisted to 15 months. There was no difference in the incidence of bleeding between the groups. Two other polymorphisms, 2677G>T/A and 1236C>T were not found to correlate with adverse events. When both the ABCB1 3435C>T and reduced function CYP2C19 allele carrier status (discussed following) were taken in to consideration, both were independent predictors of death, MI, or stroke but not stent thrombosis. When participants were grouped by both these genetic variables, 53.2% had neither, 46.8% had either, of which 18.4% had CYP2C19 allele only, 19.8% had the ABCB1 3435C>T variant only and 8.6% had both. The Kaplan–Meier MACE rates for neither variation, reduction function CYP2C19, ABCB1 3435C>T and both variations were 6.3%, 11.5%, 12.6%, and 13.6%, respectively.

The FAST MI registry also reported increased one-year adverse events (death, MI, or stroke) in ABCB1 3435TT compared with CT carriers (15.5% vs 10.7%) [12]. Conversely, Tiroch et al. examined 928 ACS PCI patients and demonstrated no difference in outcomes between CC homozygous individuals (21.8% prevalence) and T allele carriers (CC/CT; 77.2% prevalence). Other studies also observed no significant association between the ABCB1 genotype and outcomes, including the genotyping and functional testing (GIFT) substudy of the GRAVITAS trial and the East Asian MI cohort study by Jeong et al. [13–15].

In the PLATO genetic substudy, the expression of the ABCB1 gene at the nucleotide position 3435 was reported as high (CC), intermediate (CT), or low (TT) [16]. While event rates were lower with ticagrelor compared with clopidogrel in all groups, the high expressers on clopidogrel had higher MACE rates. This result in CC homozygous individuals on clopidogrel was contrary to previous data. Although no interaction was noted between treatment type and genotype, high expression of ABCB1 may impact on transport and absorption of not only clopidogrel but also ticagrelor. Consequently, a combination of genetic variations has been proposed to have a greater impact on outcomes than ABCB1 genotype alone.

Interestingly, whether aspirin has an effect on P-glycoprotein induction and reduced clopidogrel absorption has been questioned. Notably, no difference in adverse outcomes was observed with high dose (300–325 mg) versus low dose (75–100 mg) aspirin in clopidogrel-treated ACS patients in the CURRENT-OASIS 7 study [17].

CYP2C19

Nomenclature

The CYP2C19 enzyme is encoded by genes on chromosome 10 that are highly polymorphic. Several SNPs have been identified per the common-consensus star allele nomenclature from CYP2C19*2 through *10. These SNPs result in reduced function of this enzyme—thus being referred to as loss of function (LOF) alleles (Table 3.1). The CYP2C19*1 allele referred to as the wild type is associated with extensive metabolism. The *2 and *3 LOF alleles are associated with intermediate or poor metabolism. Conversely, the gain of function (GOF) allele associated with heightened response and ultrametabolizer status is the CYP2C19*17 allele. Based on combinations of these allele frequencies, the following standard nomenclature is typically used: extensive metabolizers (EM) *1/1, ultrametabolizers (UM) *1/17 or *17/17, poor metabolizers (PM) *2/2 or *3/3 or *2/3, intermediate metabolizers (IM) *1/2 or *1/3, and unknown metabolizers *2 or *3/*17.

Several pharmacodynamic studies, clinical studies and meta-analyses have examined the incidence and effect of these alleles on active metabolite levels, platelet inhibition, and ischemic and bleeding outcomes.

CYP2C19*2 and Other Loss-of-Function Alleles

The SNP 681G>A (rs4244285) in exon 5 of the long arm of chromosome 10 (10q24.1–q24.3) of the CYP2C19 gene encodes for a cryptic splice variant resulting in no enzyme activity in vivo and is referred to as the *2 allele [18].

Pharmacokinetics/Pharmacodynamics

Frere et al. studied 603 non-ST elevation myocardial infarction (NSTEMI) patients and demonstrated that noncarriers of the *2 allele had significant pharmacodynamic suppression and lower P-selectin expression compared with carriers [19]. The CYP2C19*2 polymorphism was significantly associated with ADP-induced platelet aggregation, vasodilator-stimulated phosphoprotein phosphorylation index (VASP PRI), and ADP-induced P-selectin expression in homozygous ($p<.01$, $p<.007$, and $p<.06$) and heterozygous ($p<.08$, $p<.0001$, and $p<.009$) patients compared with noncarriers; whereas CYP3A4*1B and CYP3A5*3 polymorphisms were not. The *2 allele carriers also exhibited the highest platelet reactivity on multivariate analysis ($p=.03$). This LOF allele was more frequent in clopidogrel nonresponders than responders, defined as

ADP-induced platelet aggregation >70%, $p=.03$. In a study of 196 aspirin-treated patients, CYP2C19 metabolizer status but not ABCB1 or Paraoxonase-1 (PON1) allele status influenced pharmacokinetic and pharmacodynamic measures on clopidogrel 75 mg daily [20].

Clinical Trials and Meta-analyses

The incidence of LOF alleles is variable by ethnicity and ranges from 15% in Caucasians to 45% in Pacific Islanders, 55% in Chinese, and 60.9% in East Asian individuals [13,21,22]. Several studies have demonstrated poor outcomes in patients with at least one LOF allele compared to those without [11,12,16,19,23–28]. Mega et al., compared carriers and noncarriers of LOF alleles in 1477 clopidogrel-treated patients from the TRITON TIMI 38 trial and noted that the primary ischemic endpoint of death, MI, or stroke was significantly more frequent among carriers compared with noncarriers (12.1% vs 8.0%, *p* .01) [23]. No difference was noted in the incidence of bleeding based on reduced function carrier status.

Conversely the CHARISMA, CURE, and ACTIVE A trials did not demonstrate adverse outcomes in patients with an LOF allele [29,30]. In the CHARISMA trial, 4819 patients of predominant European ancestry but only 1.4% of Asian and 4.1% of African ancestry underwent genotyping [29]. The minor allele frequencies for *2, *3, and *17 alleles were 15.4%, 0.01%, and 21.4%, respectively. Although CYP2C19 was not associated with increased ischemic events, this may be due to a stable study population of which not all had established coronary disease and only 22% underwent PCI. Furthermore the study was underpowered and neutral for the primary endpoint. However, the incidence of all GUSTO bleeding was higher in LOF noncarriers compared with carriers, suggesting reduced active metabolites in the latter.

Paré et al. investigated the influence of LOF alleles in the CURE and ACTIVE A genetic substudies [30]. In the CURE study comprising of 5059 patients, clopidogrel was better than placebo among both carriers and noncarriers of an LOF allele, without interaction. The event rates were 8.0% vs 11.6%, HR 0.69 for clopidogrel vs placebo among carriers; and 9.5% vs 13.0%, HR 0.72 among noncarriers. However, only 15% of patients in the CURE trial underwent PCI, reducing the expected benefit of clopidogrel in this trial compared with the potential benefit in a high PCI patient population. The ACTIVE A trial, comprising 1156 patients, compared DAPT with aspirin and placebo for the reduction of ischemic cerebrovascular events in patients with atrial fibrillation and was significantly underpowered for the genetic substudy to evaluate stroke outcomes. There was no interaction between genotype and study treatment for ischemic or safety outcomes.

The GIFT study tested 1028 GRAVITAS trial patients for 41 SNPs in 17 genes and found correlation between CYP2C19*2 allele and platelet aggregation at 12–24 h, 30 days, and 6 months post PCI in patients receiving 150 mg of clopidogrel, but no correlation with ABCB1 3435C>T and PON1 genes (discussed following) [14]. Compared with noncarriers, the risk of high on-treatment platelet reactivity (HTPR) 12–24h after PCI was 2.5-fold greater for carriers of one reduced-function CYP2C19 allele (*2, *3, *4, *6, and *8) and approximately 4.5-fold greater for carriers of two reduced-function alleles, irrespective of treatment assignment. The risk of HTPR at 30days attributable to reduced-function allele carriage in the patients randomly assigned to high-dose clopidogrel was 5.2% for PRU ≥ 230% and 7.3% for PRU > 208. Thus there was little incremental benefit of genotyping to identify patients at risk of HTPR late after PCI in patients treated with 150mg clopidogrel daily. In a multivariate model incorporating CYP2C19 genotype, body mass index, diabetes mellitus, age, history of congestive heart failure, and current smoking also contributed significantly to the variance in HTPR.

Mega et al. conducted a collaborative meta-analysis with study level data from nine studies, encompassing 9685 patients [27]. This included 54.5% ACS patients and 91.3% undergoing PCI. Of this cohort, 71.5% were noncarriers, 26.3% had one, and 2.2% had two CYP2C19 reduced-function alleles. The presence of any reduced function allele in clopidogrel-treated patients was a significant risk for cardiovascular events and ST. Carriers of one allele experienced increased risk of ST (HR 2.67, 95% CI 1.69–4.22, $P<.0001$) and MACE, composite of death, MI or stroke (HR 1.55, 95% CI 1.11–2.27, $P=.01$) compared with noncarriers. Similarly, in carriers of two alleles the HR for ST was 3.97 (95% CI 1.75–9.02, $P=.001$) and for MACE was 1.76 (95% CI 1.24–2.50, $P=.002$) compared with noncarriers. In landmark analyses, the HR for ST and MACE within 30days for carriers of one or two reduced-function alleles vs noncarriers was 2.94 (95% CI 1.75–4.94) and 1.36 (95% CI 1.11–1.65), respectively. Beyond 30days to the end of follow-up, the HR for carriers versus noncarriers remained high: 2.80 (95% CI 0.83–9.38) for ST and 1.61 (95% CI 0.88–2.95) for MACE.

In the meta-analysis by Hulot et al. comprising 10 studies, *2 allele carriers had an increased risk of mortality (1.8% vs 1.0%; OR 1.79, 95% CI 1.10–2.91; $n=6225$), ST (2.9% vs 0.9%, OR 3.45, 95% CI 2.14–5.57; $n=4905$) and MACE (9.7% vs 7.8%, OR 1.29, 95% CI 1.12–1.49; $n=11{,}959$) compared with noncarriers [28]. Overall at least one CYP2C19*2 allele occurred in 28% of patients.

In contrast with these meta-analyses were the findings of Holmes et al. who systematically analyzed 32 studies including six randomized studies and 42,016 patients for the effect of a LOF allele [31]. While LOF carriers had reduced active metabolite levels and reduced platelet responsiveness, there was no overall impact on composite cardiovascular outcomes of death, MI, or stroke. However, this analysis was criticized for the exclusion of studies reporting only ST, given that the latter is the most relevant endpoint for PCI patients receiving $P2Y_{12}$ receptor inhibition [32]. Furthermore, the inclusion of negative studies CHARISMA, CURE, and ACTIVE A, which included few or no PCI patients, as discussed earlier, is likely to have influenced the overall result.

In the PLATO genetic substudy, any LOF allele carriage was associated with a trend for lower 30day MACE with ticagrelor compared with clopidogrel (4.1% vs 5.7%, HR 0.73, 95% CI 0.52–1.03, $p = .078$), but not for noncarriers (3.8% vs 3.8%, HR 0.96, 95% CI 0.76–1.22, $p = .76$; interaction $p = .20$) [16]. Conversely, in landmark analysis for MACE from day 31 onward, event rates at end of follow-up were similar between treatment groups for carriers of any LOF allele (HR 0.82, 95% CI 0.58–1.17, $p = .28$), but lower for noncarriers in the ticagrelor vs clopidogrel group (HR 0.77, 95% CI 0.63–0.96, $p = .017$; interaction $p = .77$). Thus while there was an early effect of ticagrelor vs clopidogrel in carriers, this was attenuated over the course of the follow-up. Of note, PLATO included only two-thirds of patients undergoing PCI. In addition, 19.6% clopidogrel patients and 13.7% ticagrelor patients received high-dose clopidogrel loading dose ≥600mg prior to randomization.

While prior reports have indicated that high-dose clopidogrel loading can overcome the detrimental effect of reduced function alleles, this was not observed in the PLATO genetic substudy [33–35]. In patients treated with a clopidogrel loading dose of >300mg, MACE occurred in 6.6% vs 11.7% among carriers treated with ticagrelor vs clopidogrel and 5.7% vs 8.4% among noncarriers. Similarly, in patients treated with a clopidogrel loading dose of ≤300mg, the incidence of MACE was 8.8% vs 9.8% in carriers treated with ticagrelor vs clopidogrel (HR 0.89, 95% CI 0.65–1.22) and 9.1% vs 10.4% (HR 0.88, 95% CI 0.73–1.07, *p* interaction .95) in noncarriers. Notwithstanding, the impact of LOF allele genotyping was demonstrated in those undergoing an invasive vs conservative strategy. Among carriers undergoing PCI, the incidence of MACE was 7.7% vs 10.6%, HR 0.71 (0.53–0.96) with ticagrelor vs clopidogrel. Comparatively in noncarriers, the incidence of MACE was 7.4% vs 8.2%, HR 0.90 (0.73–1.10) with ticagrelor vs clopidogrel, *p* interaction .22. Conversely, in patients undergoing conservative management, the incidence of MACE was 9.6% vs 11.0% in carriers (HR 0.86, 95% CI 0.57–1.31) and 10.2% vs 11.7% (HR 0.87, 95% CI 0.69–1.11, *p* interaction .96) in noncarriers.

CYP2C19*17 Gain of Function Allele

The incidence of at least one GOF allele and ultrametabolizer status was 46.3% in a healthy Caucasian population studied by Mega et al. [23]. Compared with poor and intermediate metabolizers, ultrametabolizers had higher levels of the active metabolite as measured by the area under the concentration time curve, as well as the greatest platelet inhibition. Interestingly while there was no observable difference for active metabolite levels between UM and EM for the 75mg maintenance dose, UM achieved higher levels following 300mg or 600mg loading doses. A clear pharmacodynamic difference was also noted between UM and EM for the 600mg dose. Despite this, no clinical associations were observed for the GOF allele and MACE outcomes at

12 months in clopidogrel-treated patients of the TRITON-TIMI 38 trial [23]. Frere et al. also studied the CYP2C19*17 allele in 598 NSTEMI patients and found more intense ADP-induced platelet aggregation and lower VASP PRI in carriers vs noncarriers [36].

Similarly, Sibbing et al. confirmed these pharmacodynamic findings in 1524 PCI patients loaded with 600 mg clopidogrel. The *17 heterozygotes and homozygotes had increased platelet inhibition when tested with multiple electrode aggregometry assay, and also demonstrated an increased risk in bleeding [37]. However, no ischemic benefit was noted in terms of reduction in ST, MI, death, or composite events in *17 allele carriers—compared with extensive metabolizers (*1 homozygotes). Moreover, unlike the previous study by Mega et al., this study did not compare the *17 carriers with LOF allele carriers. In a high-risk MI PCI population of 928 patients, Tiroch et al. observed the *17 allele to be associated with reduction in ischemic events while the *2 allele and ABCB1 polymorphisms were not [15]. The primary endpoint (target lesion revascularization, TLR) showed a 37% relative reduction in carriers compared with noncarriers (14.0% vs 22.3%, $P=.002$) and a 22% reduction in a MACE composite of death, MI, and TLR (22.0% vs 28.1%, $P=.04$). Interestingly, the CURE and ACTIVE A genetic substudies also noted improved ischemic outcomes in GOF carriers (7.7% with clopidogrel vs 13.0% with placebo, HR 0.55, 95% CI 0.42–0.73) compared with noncarriers (10.0% vs 12.2%, HR 0.85, 95% CI 0.68–1.05; p interaction .02) [30].

In contradiction to these just discussed studies, the CHARISMA substudy showed neither a decrease in ischemic endpoints nor an increase in safety endpoints in patients with vs without a GOF allele, although more bleeding was noted with clopidogrel compared with placebo [29]. Conversely in the PLATO genetics trial, clopidogrel patients with a GOF allele had a higher risk of bleeding (11.9% vs 9.5%, $p=.02$) compared with patients without a GOF allele [16]. Thus reports have been conflicting for both ischemic and safety endpoints with regard to presence of a GOF allele in clopidogrel-treated patients, however a potential increase in bleeding concurrent with reduction in ischemic outcomes seems mechanistically plausible.

Incidence of CYP2C19 Alleles in Specific Ethnic Groups

In a population of 266 East Asian patients surviving acute MI, the incidence of any LOF allele was 60.9%, the highest reported in any race [13]. Conversely the incidence of the *17 allele was ~4%. In an Indian cohort study of 110 patients, the incidence of the *2 allele was 47.2% and *17 was 35.5% [38]. In the New Zealand Maori and Pacific Islander ethnic populations, the incidence of the *2 allele has been reported as 24% and 45%, respectively. Santos et al. evaluated allele frequencies in a mixed Brazilian population and identified that the incidence of the *2 allele, which was the commonest polymorphism, varied from 10.4% in Amerindians to 16.9% in Caucasians and 20.2% in persons of African descent [39]. Similarly, the incidence of the *17 allele was 15.8%, 18.0%, and 26.3%, respectively. The composite influence of the LOF and GOF alleles in different ethnic groups requires future evaluation.

CYP3A4

The CYP3A4 enzyme plays a role in the first oxidation step of clopidogrel to 2-oxo-clopidogrel. Drug interactions by substrates and inhibitors of the CYP3A4 enzyme can alter the response to clopidogrel. The CYP3A4 enzyme itself can show considerable interindividual variation, which can impact on clopidogrel metabolism and effect [40]. Lau et al. demonstrated that platelet aggregation on clopidogrel is inversely correlated with CYP3A4 activity in patients as well as healthy volunteers [41]. When coadministered with a CYP3A4 inducer like rifampicin, improved platelet response was observed in volunteers resistant to clopidogrel. Conversely, Frere et al. showed that CYP3A4 *1B or CYP3A5*3 were not associated with high platelet reactivity on clopidogrel in NSTEMI patients. Angiolillo et al. studied 82 patients in steady phase of clopidogrel therapy on a mean treatment duration of 8.6 (4.9) months, for the following CYP3A4 SNPs: *1B, *3, IVS7+258A>G, IVS7+894C>T and IVS10+12G>A [42]. Although CYP3A4*1B is generally the commonest allele variation, in this study only the last three expressed sufficient polymorphism to allow evaluation. The IVS10+12A allele carriers showed reduced activation of glycoprotein IIb/IIIA receptor and thus improved platelet responsiveness (p=.02). In contrast, in a study of 94 healthy volunteers, the intronic IVS10+12G>A polymorphism was not associated with differential ADP-induced platelet aggregation [43].

CYP1A2

The CYP1A2 enzyme that comprises 10% of the liver P450 enzymes has gained considerable attention for an enhanced clopidogrel response in smokers, thought to be mediated by induction of this enzyme [44]. In the CHARISMA study, current smokers on clopidogrel had lower all-cause and cardiovascular mortality compared with former smokers or never smokers [45]. At the same time, risk of bleeding was higher in clopidogrel randomized patients who were current smokers compared with never smokers. Park et al. studied 1431 patients on clopidogrel, of which 17% were current smokers [44]. The current smokers compared with nonsmokers had lower levels of platelet reactivity by the VerifyNow assay (227.6±76.0 vs 244.9±79.7, P=.001). Among the nine CYP1A2 SNPs studied, only the CYP1A2 (−163C>A) allele showed a genotype-dependent change in the effect of smoking on platelet reactivity. Thus not all smokers demonstrated the smokers paradox attributed to clopidogrel. Cigarette smoking was associated with an enhanced clopidogrel effect only among A allele carriers (heterozygotes and homozygotes) but not C-allele homozygotes. Smoking was also associated with a significant RRR in HTPR defined as PRU ≥235 and ≥275 among A-allele carriers but not noncarriers. While there may be conflicting evidence on whether or not clopidogrel response and bleeding risk are enhanced in smokers compared with nonsmokers [45–48], this study by Park et al. provides one explanation for the inconsistent results. Furthermore no studies have examined for racial differences in the prevalence of the CYP1A2 A-allele similar to CYP2C19 SNPs, which may provide further explanation.

Paraoxonase 1

Paraoxonase 1 (PON1) is an esterase produced in the liver that produces the active thiol metabolite by hydrolysis of 2-oxo-clopidogrel [49]. Bouman et al. suggested that PON1 rather than CYP450 was the rate-determining enzyme for the formation of the active clopidogrel metabolite in the second activation step [49]. In a microsomal ex vivo study, the authors showed that active metabolite levels were highest in PON1 RR homozygous cells and lowest in QQ homozygous cells. In a case–control study, 41 cases with nonfatal ST were compared with 71 random cases without nonfatal ST from a 7719 patient PCI cohort followed up for 18 months. PON1 QQ patients had an adjusted HR of 12.90 for ST (95% CI 4.54–95.48), whereas QR patients had HR of 4.52 (95% CI 1.81–11.24) compared with RR patients. The PON1 Q192R genotype was the only significant factor that was independently associated with the occurrence of ST in multivariate analyses ($P<.001$).

Following the administration of a 600-mg clopidogrel test dose in follow-up, patients with nonfatal ST showed a lower plasma concentration of the active clopidogrel metabolite, a higher concentration of the 2-oxo-clopidogrel metabolite, and a lower platelet inhibition compared with counterparts without ST. The PON1 Q192R polymorphism accounted for 72.5% of the variability in ADP-induced platelet aggregation after clopidogrel administration and was the primary predictor of clopidogrel response. The authors replicated these findings in a further prospective study of 1982 ACS patients with 12 month follow-up. The unadjusted HR for definite ST in QQ vs RR patients was 10.2 (95% CI, 4.39–71.43; $P<.001$). Analogously, the HR for MI was 4.93; 95% CI, 2.16–11.24; $P<0.001$ and that for MACE was 3.89; 95% CI, 2.10–7.19; $P<.001$ in QQ vs RR individuals. Concurrently, the risk of bleeding was noted to be lowest in QQ individuals (HR 0.36, 95% CI 0.18–0.75; $P=0.006$). Similarly, Nishio et al. found that patients with low PON1 activity had higher platelet reactivity by VerifyNow and higher incidence of ST [50].

In the FAST MI registry, 2210 patients including 47% NSTEMI patients treated with clopidogrel were investigated for the presence of PON1 A576G (Q192R), CYP2C19 *2 and ABCB1 3435C>T polymorphisms [51]. The incidence of the PON1 AA (QQ) SNP was 45.5%, AG (QR) was 45.4%, and GG (RR) was 9.1%. The risk of in-hospital and one-year MACE (composite of death, MI, and stroke) was similar in QQ and QR patients regardless of the presence or number of CYP2C19 LOF alleles and the ABCB1 3435C>T SNP.

However, it should be noted that the MACE in PON QQ and QR patients was not compared with RR patients.

In a pharmacodynamic study, Li et al. evaluated 180 Chinese Han patients with unstable angina for clopidogrel responsiveness in relation to PON1 QR and CYP2C19*2 allele status. Carriers of PON1 QQ had higher incidence of non-responsiveness compared with QR and RR SNPs [52]. Kang et al. determined a lower risk of bleeding in PON1 QR carriers (HR 0.61, 95% CI 0.43–0.87) [53]. They also noted that the haplotype bearing the PON1 126C allele was associated with a higher risk of CAD (OR 1.48, 95% CI 1.04–2.09, $P=.029$) and bleeding (OR 1.68, 95% CI 1.10–2.56, $P=.017$) compared with the most frequent haplotype (p-162G-126G-108T).

However, evidence for the role of PON1 QR has been conflicting with several other studies, predominantly in Chinese cohorts not showing an association with platelet reactivity or clinical outcomes [54–56]. Future studies are warranted to further examine this link and the influence on clinical treatment in the era of novel $P2Y_{12}$ agents.

CYP4F2

Arachidonic acid is metabolized to thromboxane A2 (TXA2) by the cyclo-oxygenase (COX) enzyme, which is inhibited by aspirin. In addition to TXA2, arachidonic acid is also metabolized to 20-hydroxyeicosatetraenoic (20-HETE) acid, which is produced by the enzyme CYP4F2 mono-oxygenase. HETE in turn induces oxidative stress and may potentially induce platelet reactivity. The expression of the CYP4F2 mono-oxygenase enzyme depends on the presence of inflammation but may also show genetic variation. In a study of 126 patients with ACS, the CYP4F2 *3 allele (G1347A, rs 2108622) was found to have an incidence of 45.5% in the total cohort. Platelet aggregation with epinephrine was higher in CYP4F2 GA carriers compared with GG or AA carriers [57]. However, no studies have evaluated the influence of this polymorphism in patients receiving clopidogrel or novel $P2Y_{12}$ inhibitors.

Platelet Surface Receptors

$P2Y_{12}$ Receptor

The predominant platelet inhibition by thienopyridines and ticagrelor occurs at the level of the $P2Y_{12}$ receptor mediated by ADP [57]. The $P2RY_{12}$ gene on chromosome 3 encodes this 342–amino acid receptor [58]. Fontana et al. studied the i-C139T, i-T744C, i-ins801A, and G52T polymorphisms of the $P2RY_{12}$ gene and designated H1 as the major haplotype (C in position 139, T in position 744, absence of the i-ins801A in the intron, and G in position 52 of exon 2) and H2 as the minor haplotype (T in position 139, C in position 744, presence of the i-ins801A in the intron and T in position 52 of exon 2) [58]. The respective frequencies of haplotypes H1 and H2 were 86% and 14%. The H2 haplotype was associated with increased maximal platelet aggregation in response to ADP. In another study by Staritz et al., H2 haplotype (G52T SNP) was more commonly associated with nonresponder status, and a homozygote H2 haplotype was significantly associated with clopidogrel resistance (OR 5.42) [59]. Conversely, in a study of 416 patients undergoing coronary stenting, H1/H2 haplotypes as well as the C32T SNP were not associated with a differential response after a 600-mg loading dose of clopidogrel [60]. Likewise, in a study of 119 patients, Angiolillo et al. found no effect of $P2Y_{12}$ receptor gene T744C polymorphism on platelet reactivity; however, less than one-third of these patients underwent PCI. There was no significant difference in the incidence of C-allele carriers by a 300-mg loading dose (38.9% vs 31.3%) [61]. Notably, in a study of 1378 patients, the H2 haplotype was observed in significantly more patients with CAD and among them in smokers more than nonsmokers [62]. The H2 haplotype was also associated with increased risk in patients with peripheral arterial disease through increased platelet activation [63].

These studies provide preliminary data and dedicated studies are required to explore the influence of these haplotypes on clinical outcomes in contemporary PCI.

Platelet Endothelial Aggregation Receptor-1

Platelet endothelial aggregation receptor-1 (PEAR1) is a novel membrane protein, highly expressed in platelets and endothelial cells [64]. Upon induction of platelet aggregation by physiological agonists, PEAR1 becomes phosphorylated in a glycoprotein IIb/IIIa dependent manner. The PEAR1 gene is located on chromosome 1 and the common variant A-allele is associated with increased cardiovascular outcomes in patients on DAPT undergoing PCI [65]. Lewis et al. measured platelet aggregation in 565 patients before and after DAPT from the Pharmacogenomics of Antiplatelet Intervention (PAPI) study and conducted a genome wide association study (GWAS) of drug response. Results from the GWAS revealed a strong association between SNPs on chromosome 1q23 and post-DAPT platelet aggregation. Further genotyping revealed rs12041331 in the PEAR1 gene to be most strongly associated with response to DAPT. These findings were extended in study of 227 PCI patients and 1000 patients from the INVEST GENES substudy [65]. In the PCI substudy, A-allele carriers of rs12041331 were more likely to experience a cardiovascular event or death compared with GG homozygotes (HR 2.62, 95% CI, 0.96–7.10; P=.059; and HR 3.97, 95% CI, 1.10–14.31; P=.035, respectively). In aspirin-treated INVEST-GENES patients, rs12041331 A-allele carriers had significantly increased risk of MI compared with GG homozygotes (OR 2.03, 95% confidence interval, 1.01–4.09; P=.048). In the GeneSTAR study of 1486 healthy patients in two generations of families with premature CAD, the C allele (rs2768759) was more frequent in whites (70.2%) than blacks (17.7%), and in both groups was associated with increased platelet aggregation despite aspirin treatment [66].

Similarly the intron variant G allele of rs12041331 present in 91% whites and 63% blacks was associated with increased platelet aggregation in both groups before and after aspirin treatment [67].

Protease Activated Receptors

The protease activated receptors (PAR) 1–4 on the platelet surface act as thrombin receptors [68]. The PAR-1 receptor is the target site for vorapaxar, which has been studied in both ACS and stable patients [69,70]. A meta-analysis of PAR-1 antagonists showed a significant reduction in ischemic events in patients with CAD at the expense of increased bleeding [71] The PAR-1 genotype has been found to influence platelet aggregation in CAD patients both before and during clopidogrel therapy [72]. Concurrently genetic studies on the PAR-4 receptor have shown racial differences and a higher incidence in blacks compared with whites [73]. The rs773902 SNP in the PAR-4 gene, which determines the Thr120 variant, was more common in black subjects than in white subjects (63% vs 19%) and associated with higher PAR-4–induced platelet aggregation, calcium flux, and inositol 1,4,5-triphosphate generation. Conversely, the F2RL3 variant (Phe296Val) observed only in blacks abolished the enhanced PAR-4–induced platelet aggregation and 1,4,5-triphosphate generation associated with PAR-4-Thr120. While the PAR-4 genotype did not affect vorapaxar inhibition of platelet PAR-1 function, a strong pharmacogenetic effect was observed with a PAR4-specific antagonist. These observations require future studies to examine both pharmacodynamic outcomes in larger patient numbers and clinical implications.

Nongenetic Factors

It is important to note that besides genetic characteristics, a combination of other factors can impact on clopidogrel effect, including treatment compliance, dose, drug interactions, presence of diabetes, obesity, smoking, reduced left ventricular ejection fraction, creatinine clearance, and ACS presentation [14,74].

Clopidogrel and Proton Pump Inhibitors

Clopidogrel and proton pump inhibitors (PPIs) may interact by competitive antagonism of the CYP2C19 enzyme resulting in lower levels of the active clopidogrel metabolite. However, reports on the impact of PPI coadministration on clinical outcomes have been conflicting, and whether dual administration with aspirin overcomes the effect of an interaction is unknown [75–77]. The potency for clopidogrel inhibition is strongest with lansoprazole (Ki: 0.4–1.5 mM), omeprazole (Ki: 2–6 mM), esomeprazole (Ki: 8 mM), and lower with rabeprazole (Ki: 17–21 mM) and pantoprazole (Ki: 14–69 mM) [78]. Thus pantoprazole may potentially have the lowest impact on coadministration of clopidogrel [79]. While the European Society of Cardiology (ESC) guidelines recommended prophylactic PPI in patients receiving DAPT especially in the early time period after PCI and in high-bleeding-risk patients, a position statement also provided guidelines for the appropriate choice of PPI with weaker rather than stronger CYP2C19 inhibitors [80].

The Proton Pump Inhibitors and Clopidogrel Association (PACA) trial randomized 104 NSTEMI patients to pantoprazole or omeprazole and 150 mg of clopidogrel and 75 mg of aspirin [79]. After one month, pantoprazole patients had greater platelet inhibition on VASP PRI, 36 ± 20% vs 48 ± 17% ($p = .007$), but similar inhibition with ADP aggregation. More omeprazole patients were clopidogrel nonresponders (44% vs 23%, $p = .04$), OR 2.6 (95% CI 1.2–6.2). In the Omeprazole Clopidogrel Aspirin (OCLA) double-blind placebo-controlled randomized trial, 124 PCI patients received omeprazole or placebo along with clopidogrel and aspirin [81]. On day seven VASP PRI was significantly lower in the omeprazole group, 39.8 ± 15.4% vs 51.4 ± 16.4% (p < .0001). The COGENT trial randomized 3873 patients to receive a fixed-dose combination of clopidogrel 75 mg and omeprazole 20 mg or clopidogrel and placebo in patients with an indication for DAPT [82]. While there was a significantly lower incidence of the primary gastrointestinal endpoint (overt or occult bleeding, symptomatic gastroduodenal ulcers or erosion, obstruction, or perforation) in omeprazole patients there was no difference noted in the primary cardiovascular composite of death, MI, or stroke at six months (4.9% vs 5.7%, p .96). However the study was underpowered precluding definitive conclusions in this regard. In a subanalysis of an observational genotyping study, no difference was found in platelet aggregation at 24 h in clopidogrel-treated patients receiving rabeprazole vs placebo [83]. Moreover no difference was noted in cardiovascular events at 18 months.

In 95 CAD patients, platelet reactivity was lower in CYP2C19*2 carrier patients by the VASP assay, but no difference was observed in patients treated with or without PPI. Neither CYP2C19*2 allele carriage nor PPI treatment was associated with adverse clinical outcomes [84]. Chen et al. conducted a systematic review of eight studies [85]. Compared to clopidogrel treatment alone, patients who received both a PPI and clopidogrel had higher platelet reactivity index, more ADP-induced aggregation, higher VerifyNow $P2Y_{12}$ reactivity units (PRU) (weighted mean difference of 40.58), and higher incidence of clopidogrel resistance (OR 2.49, 95% CI 1.49–4.14). However analysis of 314 patients from three randomized studies demonstrated no differences in the incidence of MACE. The incidence of adverse gastrointestinal events was reduced with PPI treatment.

In a meta-analysis of 222,311 from 23 studies, the influence of PPI on cardiovascular outcomes could not be definitively established due to significant heterogeneity between studies [86].

PRASUGREL

Metabolism

Prasugrel is a novel thienopyridine with a superior pharmacokinetic profile that demonstrates rapid, more potent, and sustained platelet inhibition compared with clopidogrel in patients with ACS [3]. In the TRITON TIMI 38 trial, prasugrel was associated with lower incidence of one-year cardiovascular composite of death, MI, or stroke as well as a markedly reduced incidence of stent thrombosis [3]. Prasugrel is hydrolyzed by esterases, in the intestine and blood, to an intermediate metabolite [87]. This undergoes activation by cytochrome P450 enzymes (mainly CYP3A and CYP2B6) to form the active metabolite R-138727, which irreversibly binds to the $P2Y_{12}$ receptor to inhibit platelet activation and aggregation (Fig. 3.1). Prasugrel metabolism thus differs from clopidogrel in the absence of a dead-end inactivation (hydrolysis) step as well as in having a single CYP activation step. However, due to the polymorphisms of CYP enzymes, interindividual variability to response can exist in both cases, despite similar affinity for the $P2Y_{12}$ receptor in vitro [26].

Pharmacokinetics, Pharmacodynamics, and Clinical Trials

Brandt et al. studied healthy subjects receiving prasugrel or clopidogrel [8]. In clopidogrel-treated subjects with the CYPC219*2 allele or CYP2C9 LOF variants, the area under the concentration time curve (AUC_{0-24}) and the maximal plasma concentration (C_{max}) was lower whereas inhibition of platelet aggregation at 4h, as well as poor responder status was significantly higher compared with noncarriers, suggesting lower exposure to the active clopidogrel metabolite. Conversely, there was no effect in carrier patients receiving prasugrel. In a recent study by Gurbel et al., CYP2C19 status did not impact on aggregation by maximum ADP platelet aggregation or VerifyNow in patients receiving prasugrel 5mg or 10mg [88]. Interestingly, a randomized study of 32 patients indicated that both high-dose clopidogrel and prasugrel were equally efficacious in reducing the incidence of HTPR in carriers of CYP2C19*2 allele [89].

Mega et al. [26] tested the pharmacokinetic and pharmacodynamic association of 50 CYP alleles in 238 healthy subjects receiving prasugrel as well as clinical outcomes in 1466 patients receiving prasugrel in the TRITON TIMI study. Carrier status for a reduced-function allele was not associated with attenuation of either the pharmacokinetic or pharmacodynamic response to prasugrel for any of the CYP genes evaluated. Moreover, even with an extended CYP2C19 genotypic classification of metabolizer status, no association was found between genotype and pharmacokinetic or pharmacodynamic response. There was also no significant efficacy or safety difference noted in prasugrel-treated patients based on genotype.

Similarly, no significant associations were noted between the ABCB1 variants and response to prasugrel or clinical outcomes, potentially an effect of rapid prasugrel metabolism [11]. A nonsignificant trend was observed for 2677TT homozygotes having higher MACE rates compared with the rest of the population, which requires future investigation.

TICAGRELOR

Metabolism

Ticagrelor is a nucleoside (cyclopentyl-triazolo-pyrimidine) analogue that is a directly acting, rapid, potent, and reversible inhibitor of the $P2Y_{12}$ platelet receptor. The CYP3A4 and CYP3A5 enzymes also mediate simple metabolism to form the active and inactive ARC metabolites, AR-C124910XX and AR-C133913XX, respectively (Fig. 3.1) [90].

Pharmacokinetics, Pharmacodynamics, and Clinical Trials

The ONSET-OFFSET study established that ticagrelor has both a rapid onset and offset of action, more profound platelet inhibition and a more persistent effect in maintenance therapy compared with clopidogrel [91]. Storey et al. also confirmed the earlier offset of ticagrelor action within 4–24h [92]. The RESPOND two-way crossover randomized trial tested the effect of ticagrelor in clopidogrel nonresponders with stable CAD [93]. Following a 300-mg clopidogrel loading dose, responders and nonresponders by light transmission aggregometry were randomized to ticagrelor (180mg loading/90mg twice daily) or clopidogrel (600mg loading/75mg daily) and all patients crossed over at 14days. Treatment with ticagrelor among nonresponders resulted in a >10%, >30%, and >50% decrease in platelet aggregation from baseline in 100%, 75%, and 13% of patients, respectively. Platelet aggregation fell from 59±9% to 35±11% in patients switched from clopidogrel to ticagrelor and increased from 36±14% to 56%±9% in patients switched from ticagrelor to clopidogrel (*p* .0001 for both).

In the ONSET/OFFSET and RESPOND genotype studies, ticagrelor exhibited lower platelet reactivity than clopidogrel by all assays (light transmission aggregometry, VerifyNow, VASP) irrespective of genotype and metabolizer status [94]. The incidence of LOF carriers was 40% in the ticagrelor arm and 28% in the clopidogrel arm (p=.10). Although the alleles were well balanced, all the GOF allele carriers fell in the ticagrelor group and this group also had more intermediate metabolizers. LOF carriers and intermediate metabolizers had higher platelet function during clopidogrel therapy, and the most influence of CYP2C19 genotype was noted during maintenance therapy compared with post loading and was most effectively demonstrated by the VerifyNow $P2Y_{12}$ assay. Platelet reactivity during clopidogrel therapy was similar in the *1/*17 and *1/*1 genotypes. Notably, the action of ticagrelor was not only superior to clopidogrel in LOF allele carriers but also in GOF and wild-type allele carriers (ultra and extensive metabolizers).

The PLATO study compared ticagrelor with clopidogrel in 18,624 patients with ACS, approximately two-thirds of which underwent PCI. Ticagrelor significantly reduced the incidence of major adverse cardiovascular events (MACE) composite of death from vascular causes, myocardial infarction, or stroke (16% RRR, 1.87% ARR) as well as a reduction in one-year all-cause mortality [2]. Although this efficacy benefit was at the expense of increase in bleeding, ticagrelor was a significant advance with respect to bypassing the pharmacokinetic shortcomings of clopidogrel. This issue is of particular importance in ST-elevation myocardial infarction and primary PCI as a shorter time to platelet inhibition would mechanistically correlate with superior outcomes. Of the PLATO study population, 10,285 patients were also genotyped [16]. Patients receiving ticagrelor had superior outcomes irrespective of the presence of loss of function alleles, as discussed previously. Furthermore, clopidogrel patients with LOF alleles had worse outcomes that those without, and patients with GOF alleles had higher bleeding than those without. The SNPs encoding for $P2Y_{12}$, $P2Y_1$, and GP3a (Integrin beta 3) receptors had no impact on the effect of ticagrelor [95]. In a GWAS, ticagrelor had a small effect on three SNPs—CYP3A4, SLCO1B1, and UBT2B7 [96]—but the clinical significance of this is not yet known.

Compliance

Compliance to medications is an important factor influencing drug bioavailability and effect. While twice-daily dosing of ticagrelor may be thought to negatively influence patient adherence, the TWICE project showed that the risk of missing two consecutive doses of ticagrelor was 26% compared with a 47% chance of clopidogrel patients missing a single daily dose. Furthermore, at this rate of nonadherence the projected platelet inhibition levels using simulations based on the ONSET/OFFSET study remained higher with ticagrelor than with clopidogrel [97].

Drug–Drug Interactions

Ticagrelor is a substrate as well as an activator and inhibitor of both hepatic and intestinal CYP3A4 enzyme [98]. Coadministration with CYP3A4 inducers like rifampicin have been shown to result in lower ticagrelor levels, lower area under the concentration–time curve, and reduced half-life from 8.4 to 2.8 h in healthy individuals. Although platelet aggregation was unaffected, the offset time was significantly shortened [99,100]. Likewise, ticagrelor may increase the availability of statins by a competing CYP3A4 mechanism [101,102]. In the PLATO trial, ticagrelor reduced all-cause mortality and vascular mortality at 30 days through to study end in patients on statins and showed a trend for lower 30 day MACE [101]. Conversely, grapefruit juice increases ticagrelor levels through intestinal metabolism [103]. In vitro studies have also shown that ticagrelor induces the CYP2B6 and CYP2C9 enzymes. However, at a therapeutic dose meaningful induction and interaction with other drugs is unlikely [90].

GENOTYPING ASSAYS

The two types of genotyping assays in contemporary practice are laboratory based and point of care. The former are time consuming and require several hours for results to be determined, which is not conducive to decision-making in clinical practice. Moreover, DNA extraction/isolation may need to be conducted separately, which is cumbersome. On the other hand, point-of-care assays are more practical and results are available within 1 h making it feasible to incorporate in clinical care. The current point-of-care assays include the Spartan (Spartan BioScience Inc, Ottawa, Ontario, Canada) and Verigene near point-of-care (Nanosphere, Northbrook, IL, USA) assays whereas the main laboratory-based assay is the TaqMan assay (Applied Biosystems, CA, USA). The Spartan device was used in the RAPID GENE proof-of-concept study, and genotyping was conducted by research nurses without need for special laboratory training [22]. The test involved four steps: obtaining a buccal swab from patients, placement of swab in the assay cartridge, insertion of the reaction solution, and activating testing with a button on the device. Results were available within 1 h. The Verigene system allows routine evaluation of SNPs, as a low-density array without PCR amplification prior to detection. Compared with the TaqMan system, Verigene has a shorter turnaround time, a higher cost of reagents, but is more cost-effective overall and practical from the perspective of personnel and resource [104]. Other available laboratory tests include the AmpliChip CYP450 (Roche Diagnostics, IN, USA), and the INFINITI Analyzer assay (Auto-Genomics, CA, USA) [105].

COST-EFFECTIVENESS OF GENOTYPING

With growing evidence on the role of genotyping, the issues of resource allocation and cost-effectiveness become increasingly relevant. Kazi et al. evaluated the cost-effectiveness of a ticagrelor for all strategy versus testing for clopidogrel responsiveness prior to prescription in ACS [106]. The authors studied five prescription strategies: generic clopidogrel, prasugrel, ticagrelor use or genotyping, and switch to ticagrelor vs clopidogrel and genotyping, and switch to prasugrel vs clopidogrel. The clopidogrel strategy produced $US 179,301 in costs and 9.428 quality adjusted life years (QALY). Genotyping with prasugrel was superior to prasugrel alone, with an incremental cost-effectiveness ratio (ICER) of $US 35,800 per QALY compared with clopidogrel. Genotyping with ticagrelor was more effective than genotyping with prasugrel ($US 30,200 per QALY relative to clopidogrel). Ticagrelor was the most effective strategy ($US 52,600 per QALY relative to genotyping with ticagrelor). Similarly an Australian study determined ticagrelor therapy in ACS to be the most cost-effective strategy compared with either genotyping and ticagrelor use or clopidogrel for all [107].

In the cost-effectiveness TRITON TIMI study, prasugrel was associated with reduced average total costs of $US 221 per patient due to reduced hospitalization over 15 months for target vessel revascularization and an increased life expectancy of 0.102 year due to decreased MI [108]. Accordingly, prasugrel was deemed an economically dominant strategy. Conversely, when this was assessed in a New Zealand national population, prasugrel for all was determined not to be cost-effective [21]. The best strategy was genotyping and prasugrel use in the ethnicities with a high incidence of the LOF alleles. Prasugrel treatment for all was cost-effective when the TRITON TIMI event rates were assumed for the local population but not when observed local event rates that were higher were considered. A genetic strategy was cost-effective for hospital observed and trial based rates ($NZ 8702/QALY vs $NZ 24,617/QALY). This was especially cost-effective in Maori ($NZ 7312/QALY) and Pacific Islander ($NZ 7041/QALY) patients who experience a higher incidence of *2 allele and higher event rates.

Despite the advantages of novel antiplatelet therapy, a subgroup of patients at higher bleeding risk will continue to warrant clopidogrel therapy. It is in this group particularly, as well as in centers with economic issues precluding novel agents in all, that genotyping may have the most important role. Furthermore in non-ACS situations, the issue of clopidogrel with geno-phenotyping versus novel therapy use requires further investigation.

GENETIC POLYMORPHISMS RELATED TO ASPIRIN

Aspirin acts by acetylating the enzymes COX1 and COX2 to inhibit production of TXA2. This prevents TXA2-mediated activation of two platelet receptors, ITGA2B and the collagen receptor ITGA2, both of which play an important role in platelet aggregation. Weng et al. studied four common polymorphisms associated with aspirin resistance, COX-1 (rs3842787, C50>T), COX-2 (rs20417, 765G>C), ITGA2B (rs20118426, 1565T>C), and ITGA2 (rs1126643, 807C>T) [109]. The greatest association in the study was found for COX2 and ITGA2 with an OR of 1.83 and 2.37, respectively, for the occurrence of aspirin resistance. However, analysis by ethnicity observed a stronger prediction of all except COX-2 polymorphisms for aspirin insensitivity in Chinese patients compared with Caucasians. On meta-regression analyses, heterogeneity for COX2 and ITGA2 was observed to be due to differences in the incidence of CAD and platelet count. Clappers et al. also studied the COX1 C50>T polymorphism and did not demonstrate an increase in clinical adverse events among carriers [110].

FUTURE DIRECTIONS

Cilostazol and Role of Genetics

Cilostazol selectively targets phosphodiesterase III (PDE-III), inhibits platelet aggregation induced by collagen, ADP, epinephrine, and arachidonic acid, as well as improves endothelial cell function [111]. By inhibiting PDE, cilostazol prevents the hydrolysis of cyclic adenosine monophosphate (cAMP) resulting in higher intracellular cAMP, inhibition of the COX pathway, and lower production of TXA2 [112]. Two major metabolites of cilostazol produced via CYP3A4 (OPC-13 015) and CYP3A5, CYP2C19 (OPC-13 213) enzymes account for 50% of PDE inhibition [113]. The use of cilostazol in CAD and the influence of genotyping has been most extensively studied in East Asian patients who despite a high incidence of CYP2C19 LOF alleles have a discordantly low risk of ST and clinical events [13].

In the ACCEL-TRIPLE study, 127 PCI patients on triple therapy with aspirin, clopidogrel, and cilostazol for at least one month were genotyped for the CYP2C19*2, *3; CYP3A4 *3 and ABCB1 3435C>T polymorphisms [114]. While the CYP3A4 and ABCB1 SNPs did not impact on platelet aggregation, the presence of CYP2C19 LOF affected platelet inhibition in a stepwise manner when tested using ADP-induced aggregation. Late platelet reactivity and PRU also varied with the presence of 0, 1 and 2 carrier alleles. On multivariate analysis, presence of two LOF alleles was a significant predictor for the prevalence of HTPR, OR 5.78.

The ACCEL-AMI study tested 126 AMI patients for use of cilostazol in addition to aspirin and standard-dose clopidogrel in patients undergoing PCI compared with double-maintenance dose of clopidogrel 150mg daily [115]. In noncarriers of the CYP2C19*2 allele, despite more potent inhibition, the incidence of HTPR did not differ between the groups. However, in carriers, platelet inhibition with 5 and 20 μm of ADP was greater with triple therapy compared with high-maintenance dose clopidogrel. Moreover the incidence of HTPR at 30 days was lower with triple therapy compared with high-dose clopidogrel (15.4% vs 44.7%, $p = .005$). Similar results were demonstrated in the genetic substudy of the CILON-T randomized trial [116].

In a pilot study of 80 diabetic patients comparing cilostazol and standard dose clopidogrel with double-dose clopidogrel, adding cilostazol achieved greater platelet inhibition with no effect from genetic polymorphisms either CYP2C19*2 or *3 alleles or the ABCB1 3535C>T polymorphism [117].

Statins and the Role of Genetics

Mega et al. studied 48,421 patients from four randomized trials (JUPITER, ASCOT, CARE, and PROVE-IT) and one cohort study (Malmo Diet and Cancer study) for 27 genetic variations [118]. A genetic score was formulated from these and the population was divided into quintiles of low, intermediate, and high genetic risk scores. A significantly increasing gradient of RRR for cardiovascular events was noted across the low, intermediate, and high-risk groups. The highest risk group had a threefold decrease in the number needed to treat (NNT) to prevent one coronary heart disease event. From the JUPITER trial, the NNT was 66, 42, and 25 respectively from low to high genetic risk groups. Further studies are needed to reproduce these results and to correlate these findings with clinical risk based on traditional cardiovascular risk factors.

SUMMARY: ROLE OF GENOTYPING IN ACUTE CORONARY SYNDROME

The overall evidence to date suggests that the commonest genetic variation with an ability to impact on clinical outcomes in ACS is the CYP2C19 *2 allele. While this and other alleles show significant racial variation, the advantage of systematically genotyping entire populations may not be cost-effective. Notwithstanding, several groups have demonstrated feasibility of incorporating genotyping in patients with and without ACS undergoing cardiac catheterization [119–122]. The feasibility of a pharmacogenomic evaluation for CYP2C19*2, CYP2C19*17, and ABCB1 TT alleles with guided anti-$P2Y_{12}$ therapy has also been studied in an STEMI population [123]. The current Clinical Pharmacogenetics Implementation Consortium guidelines recommend a potential role for genotyping in patients with ACS undergoing PCI with a proposed algorithm for care based on CYP2C19 metabolizer status [124].

Particularly in the era of novel antiplatelet therapies, genotyping requires careful consideration [125,126]. Two ongoing studies, the Patient Outcome After Primary PCI (POPular) and Tailored Antiplatelet Initiation to Lesson Outcomes Due to Decreased Clopidogrel Response After Percutaneous Coronary Intervention (TAILOR-PCI), aim to investigate the clinical outcomes of genotyping for antiplatelet therapy selection. The POPular genetics study is an open-label, multicenter trial of 2700 STEMI patients undergoing primary PCI who are randomized to CYP2C19 genotyping or routine ticagrelor or prasugrel treatment [127]. While *1/*1 (wild-type) patients will receive clopidogrel, patients carrying *2 or *3 LOF alleles will receive ticagrelor or prasugrel. The primary endpoint is one-year composite of death, recurrent MI, definite stent thrombosis, stroke, or PLATO major bleeding. The cost effectiveness of a prescription by genotyping strategy and quality-of-life assessment will also be examined. The TAILOR-PCI study will determine if genetic testing can identify the optimal antiplatelet therapy for patients undergoing PCI who may be carriers of LOF alleles (ClinicalTrials.gov Identifier NCT01742117). The primary end point is one-year MACE, a composite of nonfatal myocardial infarction, non-atal stroke, cardiovascular mortality, severe recurrent ischemia, or ST.

Until these data are forthcoming clinicians should continue to use their expert judgment and be guided by individual clinical risks for ischemia and bleeding to determine optimal prescription of clopidogrel or novel antiplatelet therapies in ACS.

REFERENCES

[1] Amsterdam EA, Wenger NK, Brindis RG, et al. 2014 AHA/ACC guideline for the management of patients with non-ST-elevation acute coronary syndromes: a report of the American College of Cardiology/American Heart Association Task Force on Practice Guidelines. J Am Coll Cardiol 2014;64(24):e139–228.

[2] Wallentin L, Becker RC, Budaj A, et al. Ticagrelor versus clopidogrel in patients with acute coronary syndromes. N Engl J Med 2009;361(11):1045–57.

[3] Wiviott SD, Braunwald E, McCabe CH, et al. Prasugrel versus clopidogrel in patients with acute coronary syndromes. N Engl J Med 2007;357(20):2001–15.

[4] Simard T,HB, Ramirez FD, Froeschl M, Chen YX, O'Brien ER. The evolution of coronary stents: a brief review. Can J Cardiol 2014;30(1):35–45.

[5] Yusuf S, Zhao F, Mehta SR, et al. Effects of clopidogrel in addition to aspirin in patients with acute coronary syndromes without ST-segment elevation. N Engl J Med 2001;345(7): 494–502.

[6] Mehta SR, Yusuf S, Peters RJ, et al. Effects of pretreatment with clopidogrel and aspirin followed by long-term therapy in patients undergoing percutaneous coronary intervention: the PCI-CURE study. Lancet 2001;358(9281):527–33.

[7] Taubert D, von Beckerath N, Grimberg G, et al. Impact of P-glycoprotein on clopidogrel absorption. Clin Pharmacol Ther 2006;80(5):486–501.

[8] Brandt JT, Close SL, Iturria SJ, et al. Common polymorphisms of CYP2C19 and CYP2C9 affect the pharmacokinetic and pharmacodynamic response to clopidogrel but not prasugrel. J Thromb Haemost 2007;5(12):2429–36.

[9] Muller I, Seyfarth M, Rudiger S, et al. Effect of a high loading dose of clopidogrel on platelet function in patients undergoing coronary stent placement. Heart 2001;85(1):92–3.

[10] Wang XQ, Shen CL, Wang BN, et al. Genetic polymorphisms of CYP2C19 2 and ABCB1 C3435T affect the pharmacokinetic and pharmacodynamic responses to clopidogrel in 401 patients with acute coronary syndrome. Gene 2015;558(2):200–7.

[11] Mega JL, Close SL, Wiviott SD, et al. Genetic variants in ABCB1 and CYP2C19 and cardiovascular outcomes after treatment with clopidogrel and prasugrel in the TRITON-TIMI 38 trial: a pharmacogenetic analysis. Lancet 2010;376(9749):1312–9.

[12] Simon T, Verstuyft C, Mary-Krause M, et al. Genetic determinants of response to clopidogrel and cardiovascular events. N Engl J Med 2009;360(4):363–75.

[13] Jeong YH, Tantry US, Kim IS, et al. Effect of CYP2C19*2 and *3 loss-of-function alleles on platelet reactivity and adverse clinical events in East Asian acute myocardial infarction survivors treated with clopidogrel and aspirin. Circ Cardiovasc Interv 2011;4(6):585–94.

[14] Price MJ, Murray SS, Angiolillo DJ, et al. Influence of genetic polymorphisms on the effect of high- and standard-dose clopidogrel after percutaneous coronary intervention: the GIFT (Genotype Information and Functional Testing) study. J Am Coll Cardiol 2012;59(22):1928–37.

[15] Tiroch KA, Sibbing D, Koch W, et al. Protective effect of the CYP2C19 *17 polymorphism with increased activation of clopidogrel on cardiovascular events. Am Heart J 2010;160(3):506–12.

[16] Wallentin L, James S, Storey RF, et al. Effect of CYP2C19 and ABCB1 single nucleotide polymorphisms on outcomes of treatment with ticagrelor versus clopidogrel for acute coronary syndromes: a genetic substudy of the PLATO trial. Lancet 2010;376(9749):1320–8.

[17] Wessler JD, Grip LT, Mendell J, Giugliano RP. The P-glycoprotein transport system and cardiovascular drugs. J Am Coll Cardiol 2013;61(25):2495–502.

[18] Trenk D, Hochholzer W, Fromm MF, et al. Cytochrome P450 2C19 681G>A polymorphism and high on-clopidogrel platelet reactivity associated with adverse 1-year clinical outcome of elective percutaneous coronary intervention with drug-eluting or bare-metal stents. J Am Coll Cardiol 2008;51(20):1925–34.

[19] Frere C, Cuisset T, Morange PE, et al. Effect of cytochrome p450 polymorphisms on platelet reactivity after treatment with clopidogrel in acute coronary syndrome. Am J Cardiol 2008;101(8):1088–93.

[20] Braun OO, Angiolillo DJ, Ferreiro JL, et al. Enhanced active metabolite generation and platelet inhibition with prasugrel compared to clopidogrel regardless of genotype in thienopyridine metabolic pathways. Thromb Haemost 2013;110(6):1223–31.

[21] Panattoni L, Brown PM, Te Ao B, Webster M, Gladding P. The cost effectiveness of genetic testing for CYP2C19 variants to guide thienopyridine treatment in patients with acute coronary syndromes: a New Zealand evaluation. Pharmacoeconomics 2012;30(11): 1067–84.

[22] Roberts JD, Wells GA, Le May MR, et al. Point-of-care genetic testing for personalisation of antiplatelet treatment (RAPID GENE): a prospective, randomised, proof-of-concept trial. Lancet 2012;379(9827):1705–11.

[23] Mega JL, Close SL, Wiviott SD, et al. Cytochrome p-450 polymorphisms and response to clopidogrel. N Engl J Med 2009;360(4):354–62.

[24] Sibbing D, Stegherr J, Latz W, et al. Cytochrome P450 2C19 loss-of-function polymorphism and stent thrombosis following percutaneous coronary intervention. Eur Heart J 2009;30(8):916–22.

[25] Collet JP, Hulot JS, Pena A, et al. Cytochrome P450 2C19 polymorphism in young patients treated with clopidogrel after myocardial infarction: a cohort study. Lancet 2009;373(9660):309–17.

[26] Mega JL, Close SL, Wiviott SD, et al. Cytochrome P450 genetic polymorphisms and the response to prasugrel: relationship to pharmacokinetic, pharmacodynamic, and clinical outcomes. Circulation 2009;119(19):2553–60.

[27] Mega JL, Simon T, Collet JP, et al. Reduced-function CYP2C19 genotype and risk of adverse clinical outcomes among patients treated with clopidogrel predominantly for PCI: a meta-analysis. JAMA 2010;304(16):1821–30.

[28] Hulot JS, Collet JP, Silvain J, et al. Cardiovascular risk in clopidogrel-treated patients according to cytochrome P450 2C19*2 loss-of-function allele or proton pump inhibitor coadministration: a systematic meta-analysis. J Am Coll Cardiol 2010;56(2):134–43.

[29] Bhatt DL, Pare G, Eikelboom JW, et al. The relationship between CYP2C19 polymorphisms and ischaemic and bleeding outcomes in stable outpatients: the CHARISMA genetics study. Eur Heart J 2012;33(17):2143–50.

[30] Pare G, Mehta SR, Yusuf S, et al. Effects of CYP2C19 genotype on outcomes of clopidogrel treatment. N Engl J Med 2010;363(18):1704–14.

[31] Holmes MV, Perel P, Shah T, Hingorani AD, Casas JP. CYP2C19 genotype, clopidogrel metabolism, platelet function, and cardiovascular events: a systematic review and meta-analysis. JAMA 2011;306(24):2704–14.

[32] Mega JL, Topol EJ, Sabatine MS. CYP2C19 genotype and cardiovascular events. JAMA 2012;307(14):1482–3. Author reply 1484–5.

[33] Hulot JS, Wuerzner G, Bachelot-Loza C, et al. Effect of an increased clopidogrel maintenance dose or lansoprazole co-administration on the antiplatelet response to clopidogrel in CYP2C19-genotyped healthy subjects. J Thromb Haemost 2010;8(3):610–3.

[34] Jeong YH, Kim IS, Park Y, et al. Carriage of cytochrome 2C19 polymorphism is associated with risk of high post-treatment platelet reactivity on high maintenance-dose clopidogrel of 150 mg/day: results of the ACCEL-DOUBLE (Accelerated Platelet Inhibition by a Double Dose of Clopidogrel According to Gene Polymorphism) study. JACC Cardiovasc Interv 2010;3(7):731–41.
[35] Hulot JS, Collet JP, Montalescot G. Genetic substudy of the PLATO trial. Lancet 2011;377(9766):637. Author reply 637–8.
[36] Frere C, Cuisset T, Gaborit B, Alessi MC, Hulot JS. The CYP2C19*17 allele is associated with better platelet response to clopidogrel in patients admitted for non-ST acute coronary syndrome. J Thromb Haemost 2009;7(8):1409–11.
[37] Sibbing D, Koch W, Gebhard D, et al. Cytochrome 2C19*17 allelic variant, platelet aggregation, bleeding events, and stent thrombosis in clopidogrel-treated patients with coronary stent placement. Circulation 2010;121(4):512–8.
[38] Shetkar SS, Ramakrishnan S, Seth S, et al. CYP 450 2C19 polymorphisms in Indian patients with coronary artery disease. Indian Heart J 2014;66(1):16–24.
[39] Santos PC, Soares RA, Santos DB, et al. CYP2C19 and ABCB1 gene polymorphisms are differently distributed according to ethnicity in the Brazilian general population. BMC Med Genet 2011;12:13.
[40] Lamba JK, Lin YS, Schuetz EG, Thummel KE. Genetic contribution to variable human CYP3A-mediated metabolism. Adv Drug Deliv Rev 2002;54(10):1271–94.
[41] Lau WC, Gurbel PA, Watkins PB, et al. Contribution of hepatic cytochrome P450 3A4 metabolic activity to the phenomenon of clopidogrel resistance. Circulation 2004;109(2):166–71.
[42] Angiolillo DJ, Fernandez-Ortiz A, Bernardo E, et al. Contribution of gene sequence variations of the hepatic cytochrome P450 3A4 enzyme to variability in individual responsiveness to clopidogrel. Arterioscler Thromb Vasc Biol 2006;26(8):1895–900.
[43] Fontana P, Hulot JS, De Moerloose P, Gaussem P. Influence of CYP2C19 and CYP3A4 gene polymorphisms on clopidogrel responsiveness in healthy subjects. J Thromb Haemost 2007;5(10):2153–5.
[44] Park KW, Park JJ, Jeon KH, et al. Enhanced clopidogrel responsiveness in smokers: smokers' paradox is dependent on cytochrome P450 CYP1A2 status. Arterioscler Thromb Vasc Biol 2011;31(3):665–71.
[45] Berger JS, Bhatt DL, Steinhubl SR, et al. Smoking, clopidogrel, and mortality in patients with established cardiovascular disease. Circulation 2009;120(23):2337–44.
[46] Sibbald M, Yan AT, Huang W, et al. Association between smoking, outcomes, and early clopidogrel use in patients with acute coronary syndrome: insights from the Global Registry of Acute Coronary Events. Am Heart J 2010;160(5):855–61.
[47] Bliden KP, Dichiara J, Lawal L, et al. The association of cigarette smoking with enhanced platelet inhibition by clopidogrel. J Am Coll Cardiol 2008;52(7):531–3.
[48] Gremmel T, Steiner S, Seidinger D, et al. Smoking promotes clopidogrel-mediated platelet inhibition in patients receiving dual antiplatelet therapy. Thromb Res 2009;124(5):588–91.
[49] Bouman HJ, Schomig E, van Werkum JW, et al. Paraoxonase-1 is a major determinant of clopidogrel efficacy. Nat Med 2011;17(1):110–6.
[50] Nishio R, Shinke T, Otake H, et al. Paraoxonase-1 activity affects the clopidogrel response in CYP2C19 loss-of-function carriers. Thromb Res 2013;132(5):558–64.
[51] Simon T, Steg PG, Becquemont L, et al. Effect of paraoxonase-1 polymorphism on clinical outcomes in patients treated with clopidogrel after an acute myocardial infarction. Clin Pharmacol Ther 2011;90(4):561–7.

[52] Li X, Zhang L, Chen X, et al. PON1 Q192R genotype influences clopidogrel responsiveness by relative platelet inhibition instead of on-treatment platelet reactivity. Thromb Res 2013;132(4):444–9.
[53] Kang YH, Lao HY, Wu H, et al. Association of PON1 genotype and haplotype with susceptibility to coronary artery disease and clinical outcomes in dual antiplatelet-treated Han Chinese patients. Eur J Clin Pharmacol 2013;69(8):1511–9.
[54] Zhang L, Chen Y, Jin Y, et al. Genetic determinants of high on-treatment platelet reactivity in clopidogrel treated Chinese patients. Thromb Res 2013;132(1):81–7.
[55] Tang XF, Wang J, Zhang JH, et al. Effect of the CYP2C19 2 and 3 genotypes, ABCB1 C3435T and PON1 Q192R alleles on the pharmacodynamics and adverse clinical events of clopidogrel in Chinese people after percutaneous coronary intervention. Eur J Clin Pharmacol 2013;69(5):1103–12.
[56] Chan MY, Tan K, Tan HC, et al. CYP2C19 and PON1 polymorphisms regulating clopidogrel bioactivation in Chinese, Malay and Indian subjects. Pharmacogenomics 2012;13(5):533–42.
[57] Tatarunas V, Jankauskiene L, Kupstyte N, et al. The role of clinical parameters and of CYP2C19 G681 and CYP4F2 G1347A polymorphisms on platelet reactivity during dual antiplatelet therapy. Blood Coagul Fibrinolysis 2014;25(4):369–74.
[58] Fontana P, Dupont A, Gandrille S, et al. Adenosine diphosphate-induced platelet aggregation is associated with P2Y12 gene sequence variations in healthy subjects. Circulation 2003;108(8):989–95.
[59] Staritz P, Kurz K, Stoll M, et al. Platelet reactivity and clopidogrel resistance are associated with the H2 haplotype of the P2Y12-ADP receptor gene. Int J Cardiol 2009;133(3):341–5.
[60] von Beckerath N, von Beckerath O, Koch W, et al. P2Y12 gene H2 haplotype is not associated with increased adenosine diphosphate-induced platelet aggregation after initiation of clopidogrel therapy with a high loading dose. Blood Coagul Fibrinolysis 2005;16(3): 199–204.
[61] Angiolillo DJ, Fernandez-Ortiz A, Bernardo E, et al. Lack of association between the P2Y12 receptor gene polymorphism and platelet response to clopidogrel in patients with coronary artery disease. Thromb Res 2005;116(6):491–7.
[62] Cavallari U, Trabetti E, Malerba G, et al. Gene sequence variations of the platelet P2Y12 receptor are associated with coronary artery disease. BMC Med Genet 2007;8:59.
[63] Fontana P, Gaussem P, Aiach M, et al. P2Y12 H2 haplotype is associated with peripheral arterial disease: a case-control study. Circulation 2003;108(24):2971–3.
[64] Nanda N, Bao M, Lin H, et al. Platelet endothelial aggregation receptor 1 (PEAR1), a novel epidermal growth factor repeat-containing transmembrane receptor, participates in platelet contact-induced activation. J Biol Chem 2005;280(26):24680–9.
[65] Lewis JP, Ryan K, O'Connell JR, et al. Genetic variation in PEAR1 is associated with platelet aggregation and cardiovascular outcomes. Circ Cardiovasc Genet 2013;6(2):184–92.
[66] Herrera-Galeano JE, Becker DM, Wilson AF, et al. A novel variant in the platelet endothelial aggregation receptor-1 gene is associated with increased platelet aggregability. Arterioscler Thromb Vasc Biol 2008;28(8):1484–90.
[67] Williams MS, Weiss EJ, Sabatine MS, et al. Genetic regulation of platelet receptor expression and function: application in clinical practice and drug development. Arterioscler Thromb Vasc Biol 2010;30(12):2372–84.
[68] Steinberg SF. The cardiovascular actions of protease-activated receptors. Mol Pharmacol 2005;67(1):2–11.
[69] Tricoci P, Huang Z, Held C, et al. Thrombin-receptor antagonist vorapaxar in acute coronary syndromes. N Engl J Med 2012;366(1):20–33.

[70] Bonaca MP, Scirica BM, Braunwald E, et al. Coronary stent thrombosis with vorapaxar versus placebo: results from the TRA 2 degrees P-TIMI 50 trial. J Am Coll Cardiol 2014;64(22):2309–17.
[71] Capodanno D, Bhatt DL, Goto S, et al. Safety and efficacy of protease-activated receptor-1 antagonists in patients with coronary artery disease: a meta-analysis of randomized clinical trials. J Thromb Haemost 2012;10(10):2006–15.
[72] Smith SM, Judge HM, Peters G, et al. PAR-1 genotype influences platelet aggregation and procoagulant responses in patients with coronary artery disease prior to and during clopidogrel therapy. Platelets 2005;16(6):340–5.
[73] Edelstein LC, Simon LM, Lindsay CR, et al. Common variants in the human platelet PAR4 thrombin receptor alter platelet function and differ by race. Blood 2014;124(23):3450–8.
[74] Liang Y, Hirsh J, Weitz JI, et al. Active metabolite concentration of clopidogrel in patients taking different doses of aspirin: results of the interaction trial. J Thromb Haemost 2015;13(3):347–52.
[75] Dunn SP, Steinhubl SR, Bauer D, et al. Impact of proton pump inhibitor therapy on the efficacy of clopidogrel in the CAPRIE and CREDO trials. J Am Heart Assoc 2013;2(1):e004564.
[76] O'Donoghue ML, Braunwald E, Antman EM, et al. Pharmacodynamic effect and clinical efficacy of clopidogrel and prasugrel with or without a proton-pump inhibitor: an analysis of two randomised trials. Lancet 2009;374(9694):989–97.
[77] Goodman SG, Clare R, Pieper KS, et al. Association of proton pump inhibitor use on cardiovascular outcomes with clopidogrel and ticagrelor: insights from the platelet inhibition and patient outcomes trial. Circulation 2012;125(8):978–86.
[78] Li XQ, Andersson TB, Ahlstrom M, Weidolf L. Comparison of inhibitory effects of the proton pump-inhibiting drugs omeprazole, esomeprazole, lansoprazole, pantoprazole, and rabeprazole on human cytochrome P450 activities. Drug Metab Dispos 2004;32(8):821–7.
[79] Cuisset T, Frere C, Quilici J, et al. Comparison of omeprazole and pantoprazole influence on a high 150-mg clopidogrel maintenance dose the PACA (Proton Pump Inhibitors and Clopidogrel Association) prospective randomized study. J Am Coll Cardiol 2009;54(13):1149–53.
[80] Agewall S, Cattaneo M, Collet JP, et al. Expert position paper on the use of proton pump inhibitors in patients with cardiovascular disease and antithrombotic therapy. Eur Heart J 2013;34(23):1708–13. 1713a–1713b.
[81] Gilard M, Arnaud B, Cornily JC, et al. Influence of omeprazole on the antiplatelet action of clopidogrel associated with aspirin: the randomized, double-blind OCLA (Omeprazole CLopidogrel Aspirin) study. J Am Coll Cardiol 2008;51(3):256–60.
[82] Bhatt DL, Cryer BL, Contant CF, et al. Clopidogrel with or without omeprazole in coronary artery disease. N Engl J Med 2010;363(20):1909–17.
[83] Hokimoto S, Mizobe M, Akasaka T, et al. Impact of CYP2C19 polymorphism and proton pump inhibitors on platelet reactivity to clopidogrel and clinical outcomes following stent implantation. Thromb Res 2014;133(4):599–605.
[84] Tsantes AE, Ikonomidis I, Papadakis I, et al. Impact of the proton pump inhibitors and CYP2C19*2 polymorphism on platelet response to clopidogrel as assessed by four platelet function assays. Thromb Res 2013;132(2):e105–11.
[85] Chen J, Chen SY, Lian JJ, Zeng XQ, Luo TC. Pharmacodynamic impacts of proton pump inhibitors on the efficacy of clopidogrel in vivo–a systematic review. Clin Cardiol 2013;36(4):184–9.
[86] Kwok CS, Jeevanantham V, Dawn B, Loke YK. No consistent evidence of differential cardiovascular risk amongst proton-pump inhibitors when used with clopidogrel: meta-analysis. Int J Cardiol 2013;167(3):965–74.

[87] Jakubowski JA, Winters KJ, Naganuma H, Wallentin L. Prasugrel: a novel thienopyridine antiplatelet agent. A review of preclinical and clinical studies and the mechanistic basis for its distinct antiplatelet profile. Cardiovasc Drug Rev 2007;25(4):357–74.
[88] Gurbel PA, Bergmeijer TO, Tantry US, et al. The effect of CYP2C19 gene polymorphisms on the pharmacokinetics and pharmacodynamics of prasugrel 5-mg, prasugrel 10-mg and clopidogrel 75-mg in patients with coronary artery disease. Thromb Haemost 2014;112(3): 589–97.
[89] Sardella G, Calcagno S, Mancone M, et al. Pharmacodynamic effect of switching therapy in patients with high on-treatment platelet reactivity and genotype variation with high clopidogrel Dose versus prasugrel: the RESET GENE trial. Circ Cardiovasc Interv 2012;5(5): 698–704.
[90] Zhou D, Andersson TB, Grimm SW. In vitro evaluation of potential drug-drug interactions with ticagrelor: cytochrome P450 reaction phenotyping, inhibition, induction, and differential kinetics. Drug Metab Dispos 2011;39(4):703–10.
[91] Gurbel PA, Bliden KP, Butler K, et al. Randomized double-blind assessment of the ONSET and OFFSET of the antiplatelet effects of ticagrelor versus clopidogrel in patients with stable coronary artery disease: the ONSET/OFFSET study. Circulation 2009;120(25):2577–85.
[92] Storey RF, Bliden KP, Ecob R, et al. Earlier recovery of platelet function after discontinuation of treatment with ticagrelor compared with clopidogrel in patients with high antiplatelet responses. J Thromb Haemost 2011;9(9):1730–7.
[93] Gurbel PA, Bliden KP, Butler K, et al. Response to ticagrelor in clopidogrel nonresponders and responders and effect of switching therapies: the RESPOND study. Circulation 2010;121(10):1188–99.
[94] Tantry US, Bliden KP, Wei C, et al. First analysis of the relation between CYP2C19 genotype and pharmacodynamics in patients treated with ticagrelor versus clopidogrel: the ONSET/OFFSET and RESPOND genotype studies. Circ Cardiovasc Genet 2010;3(6):556–66.
[95] Storey RF, Melissa Thornton S, Lawrance R, et al. Ticagrelor yields consistent dose-dependent inhibition of ADP-induced platelet aggregation in patients with atherosclerotic disease regardless of genotypic variations in P2RY12, P2RY1, and ITGB3. Platelets 2009;20(5): 341–8.
[96] Varenhorst C, Eriksson N, Johansson A, et al. Effect of genetic variations on ticagrelor plasma levels and clinical outcomes. Eur Heart J 2015.
[97] Vrijens B, Claeys MJ, Legrand V, Vandendriessche E, Van de Werf F. Projected inhibition of platelet aggregation with ticagrelor twice daily vs clopidogrel once daily based on patient adherence data (the TWICE project). Br J Clin Pharmacol 2014;77(5):746–55.
[98] Teng R, Butler K. The effect of ticagrelor on the metabolism of midazolam in healthy volunteers. Clin Ther 2013;35(7):1025–37.
[99] Teng R, Mitchell P, Butler K. Effect of rifampicin on the pharmacokinetics and pharmacodynamics of ticagrelor in healthy subjects. Eur J Clin Pharmacol 2013;69(4):877–83.
[100] Weeks P, Sieg A, Vahdat K, Raissi F, Nathan S. Improved ticagrelor antiplatelet effect on discontinuation of phenytoin. Ann Pharmacother 2014;48(5):644–7.
[101] Dinicolantonio JJ, Serebruany VL. Exploring the ticagrelor-statin interplay in the PLATO trial. Cardiology 2013;124(2):105–7.
[102] Ferri N, Corsini A, Bellosta S. Pharmacology of the new P2Y12 receptor inhibitors: insights on pharmacokinetic and pharmacodynamic properties. Drugs 2013;73(15):1681–709.
[103] Holmberg MT, Tornio A, Joutsi-Korhonen L, et al. Grapefruit juice markedly increases the plasma concentrations and antiplatelet effects of ticagrelor in healthy subjects. Br J Clin Pharmacol 2013;75(6):1488–96.

[104] Saracini C, Vestrini A, Galora S, et al. Pharmacogenetics of clopidogrel: comparison between a standard and a rapid genetic testing. Genet Test Mol Biomarkers 2012;16(6):500–3.

[105] Tantry US, Jeong YH, Navarese EP, Kubica J, Gurbel PA. Influence of genetic polymorphisms on platelet function, response to antiplatelet drugs and clinical outcomes in patients with coronary artery disease. Expert Rev Cardiovasc Ther 2013;11(4):447–62.

[106] Kazi DS, Garber AM, Shah RU, et al. Cost-effectiveness of genotype-guided and dual antiplatelet therapies in acute coronary syndrome. Ann Intern Med 2014;160(4):221–32.

[107] Sorich MJ, Horowitz JD, Sorich W, et al. Cost-effectiveness of using CYP2C19 genotype to guide selection of clopidogrel or ticagrelor in Australia. Pharmacogenomics 2013;14(16):2013–21.

[108] Mahoney EM, Wang K, Arnold SV, et al. Cost-effectiveness of prasugrel versus clopidogrel in patients with acute coronary syndromes and planned percutaneous coronary intervention: results from the trial to assess improvement in therapeutic outcomes by optimizing platelet inhibition with prasugrel-thrombolysis in myocardial infarction TRITON-TIMI 38. Circulation 2010;121(1):71–9.

[109] Weng Z, Li X, Li Y, et al. The association of four common polymorphisms from four candidate genes (COX-1, COX-2, ITGA2B, ITGA2) with aspirin insensitivity: a meta-analysis. PLoS One 2013;8(11):e78093.

[110] Clappers N, van Oijen MG, Sundaresan S, et al. The C50T polymorphism of the cyclooxygenase-1 gene and the risk of thrombotic events during low-dose therapy with acetyl salicylic acid. Thromb Haemost 2008;100(1):70–5.

[111] Goto S. Cilostazol: potential mechanism of action for antithrombotic effects accompanied by a low rate of bleeding. Atheroscler Suppl 2005;6(4):3–11.

[112] Cleanthis M, Bhattacharya V, Smout J, Ashour H, Stansby G. Combined aspirin and cilostazol treatment is associated with reduced platelet aggregation and prevention of exercise-induced platelet activation. Eur J Vasc Endovasc Surg 2009;37(5):604–10.

[113] Jeong YH, Park Y, Muse WC, et al. Pharmacodynamic effect of clopidogrel therapy and switching to cilostazol in patients with the CYP2C19 loss-of-function allele (ACCEL-SWITCH) study. J Thromb Haemost 2012;10(8):1685–8.

[114] Kim IS, Jeong YH, Park Y, et al. Interaction analysis between genetic polymorphisms and pharmacodynamic effect in patients treated with adjunctive cilostazol to dual antiplatelet therapy: results of the ACCEL-TRIPLE (Accelerated Platelet Inhibition by Triple Antiplatelet Therapy According to Gene Polymorphism) study. Br J Clin Pharmacol 2012;73(4):629–40.

[115] Kim IS, Jeong YH, Park Y, et al. Platelet inhibition by adjunctive cilostazol versus high maintenance-dose clopidogrel in patients with acute myocardial infarction according to cytochrome P450 2C19 genotype. JACC Cardiovasc Interv 2011;4(4):381–91.

[116] Park KW, Park JJ, Lee SP, et al. Cilostazol attenuates on-treatment platelet reactivity in patients with CYP2C19 loss of function alleles receiving dual antiplatelet therapy: a genetic substudy of the CILON-T randomised controlled trial. Heart 2011;97(8):641–7.

[117] Jeong YH, Tantry US, Park Y, et al. Pharmacodynamic effect of cilostazol plus standard clopidogrel versus double-dose clopidogrel in patients with type 2 diabetes undergoing percutaneous coronary intervention. Diabetes Care 2012;35(11):2194–7.

[118] Mega JL, Stitziel NO, Smith JG, et al. Genetic risk, coronary heart disease events, and the clinical benefit of statin therapy: an analysis of primary and secondary prevention trials. Lancet 2015.

[119] Lee JA, Lee CR, Reed BN, et al. Implementation and evaluation of a CYP2C19 genotype-guided antiplatelet therapy algorithm in high-risk coronary artery disease patients. Pharmacogenomics 2015;16(4):303–13.

[120] Xie X, Ma YT, Yang YN, et al. Personalized antiplatelet therapy according to CYP2C19 genotype after percutaneous coronary intervention: a randomized control trial. Int J Cardiol 2013;168(4):3736–40.

[121] Ahn SG, Yoon J, Kim J, et al. Genotype- and phenotype-directed personalization of antiplatelet treatment in patients with non-ST elevation acute coronary syndromes undergoing coronary stenting. Korean Circ J 2013;43(8):541–9.

[122] Stimpfle F, Karathanos A, Droppa M, et al. Impact of point-of-care testing for CYP2C19 on platelet inhibition in patients with acute coronary syndrome and early dual antiplatelet therapy in the emergency setting. Thromb Res 2014;134(1):105–10.

[123] So DY, Wells GA, McPherson R, et al. A prospective randomized evaluation of a pharmacogenomic approach to antiplatelet therapy among patients with ST-elevation myocardial infarction: the RAPID STEMI study. Pharmacogenomics J 2015.

[124] Scott SA, Sangkuhl K, Stein CM, et al. Clinical Pharmacogenetics Implementation Consortium guidelines for CYP2C19 genotype and clopidogrel therapy: 2013 update. Clin Pharmacol Ther 2013;94(3):317–23.

[125] Kassimis G, Stavrou EF, Alexopoulos D, Athanassiadou A. Cytochrome P450 2C19 polymorphism and antiplatelet therapy. Who should really be genotyped? Curr Pharm Des 2013;19(13):2489–95.

[126] Chan NC, Eikelboom JW, Ginsberg JS, et al. Role of phenotypic and genetic testing in managing clopidogrel therapy. Blood 2014;124(5):689–99.

[127] Bergmeijer TO, Janssen PW, Schipper JC, et al. CYP2C19 genotype-guided antiplatelet therapy in ST-segment elevation myocardial infarction patients-Rationale and design of the Patient Outcome after primary PCI (POPular) Genetics study. Am Heart J 2014;168(1): 16–22 e1.

Chapter 4

Global Epidemiology and Incidence of Cardiovascular Disease

C.J. McAloon[1], F. Osman[1], P. Glennon[1], P.B. Lim[2], S.A. Hayat[1]
[1]*University Hospital Coventry, Coventry, United Kingdom;* [2]*Imperial College NHS Health Care Trust, London, United Kingdom*

List of Abbreviations

AF Atrial fibrillation
CHD Congenital heart disease
CVD Cardiovascular disease
DALY Disability-adjusted life year
HF Heart failure
HI High income
HIS Human immunodeficiency virus
ICD International classification of diseases
IHD Ischemic heart disease
LMI Low-middle income
NCD Noncommunicable diseases
WHO World Health Organization
YLD Years lived with disability
YLL Years of life lost

INTRODUCTION

The pattern and global burden of disease has evolved considerably over the last two decades from primarily communicable, maternal, and perinatal causes to noncommunicable disease (NCD) [1,2]. Cardiovascular disease (CVD) has become the single most important and largest cause of NCD deaths worldwide, at over 50% [1,3]. The World Health Organization (WHO) estimates that 17.6 million people died of CVD worldwide in 2012 [4]. Proportionally this accounts for an estimated 31.3% of global mortality, with ischemic heart disease (IHD) accounting for 7.4 million deaths, 13.2% [4]. IHD was also the greatest single cause of death in 2000, accounting for an estimated 6.0 million

Cardiovascular Diseases.

deaths. The global burden of CVD falls, principally, on the lower- and middle-income countries, accounting for over 80% of CVD deaths [5].

Individual populations face differing health challenges, and each specific population has unique health burdens, however, CVD continues to remain one of the greatest health challenges worldwide. The specific pattern and severity of disease burdens can vary greatly between populations, regions, genders, age, and ethnic groups. One of the greatest factors influencing variations in global health and CVD burden is prosperity. The differences in CVD burden are most apparent when comparing Sub-Saharan Africa to other regions across the world. In Sub-Saharan Africa CVD burden is relatively low compared to other regions of the world [6].

Cardiovascular epidemiological studies have demonstrated the association of many risk factors with the development of CVD. Exposure to certain risk factors within specific populations drives the local CVD health burden and contributes to the worldwide burden. Understanding the exposure and prevalence of these risk factors, both modifiable and nonmodifiable, will allow health care professionals to continue to perform public health interventions to modify these CVD risk factors and reduce the far-reaching effects of the associated morbidity and mortality [1]. Although smoking tobacco is a worldwide problem, prevalence is highest in Eastern Europe, Russia, and Southeast Asia. This is clearly reflective of the specific CVD present and its burden in these regions. Recognition of the prevalence of CVD risk factors is fundamental to understanding the pattern and burden, both regionally and globally.

This chapter examines current cardiovascular epidemiology, utilizing CVD data from the WHO/World Bank and the Global Burden of Disease study, which is reviewed and presented to examine the latest patterns of CVD disease globally [4,7–12]. Specific analysis is performed on fatal and nonfatal CVD burden. Furthermore, risk factor exposure and prevalence is examined to determine explanations for the current pattern of CVD disease. Specific CVD categories and conditions are also examined to elucidate a more definitive analysis on global CVD disease determinants.

GLOBAL BURDEN OF DISEASE: REPORTING METRICS

There are several reporting metrics of CVD global burden used to define fatal and nonfatal CVD. A standard definition of health conditions must be applied globally for these specific metrics to be comparable to each other. The International Classification of Diseases Tenth Edition (ICD-10) is the standard diagnostic tool for epidemiology, health management, and clinical purposes [13]. All WHO member states use the ICD-10, which has been translated into 43 different languages [13]. The ICD-10 is used to code all health diagnoses, and most member states use it to define primary mortality diagnosis.

The specific CVD conditions that are examined in the WHO global health estimates 2014 and the Global Burden of Disease Study 2010 (GBD) are rheumatic heart disease, ischemic heart disease, hypertensive heart disease, stroke, cardiomyopathy, myocarditis, endocarditis, and other circulatory diseases [4,9,14]. Congenital heart anomalies is categorized separately under congenital anomalies and does not contribute to the CVD numbers [4,9,14]. The classification of conditions is based on ICD-10, which provides a measure of standardization in capturing the same CVD conditions globally. However, several conditions are not specifically coded for by ICD-10 and thus not easily captured [13]. Heart failure is a common condition that is not directly coded for by 1CD-10, leading to issues in establishing its prevalence internationally [2].

Several metrics have been used to quantify morbidity and mortality. Mortality rate is a critically important measure of health status and disease burden. However, a mortality rate assigns the same value to a death no matter the age it occurred. Applying a weighting to mortality to account for prematurity of death allows the number of predicted years for that life lost to be accounted for [1]. This metric is called "year of life lost" (YLL) and takes into account both the frequency and age of deaths. YLL is calculated by subtracting the age of death from the highest possible life expectancy for a person of that age [15].

This metric focuses on mortality but does not account for those members of the population that have been impacted with nonfatal disease and have had less than ideal health. The "years lived with disability" (YLD) metric accounts for years lost from ideal health status. YLD is measured by taking the prevalence of the condition multiplied by the disability weight for that condition. Disability weights are provided by the estimation of the severity of that particular condition [15]. The best measure of total disease impact is the "disability adjusted life year" (DALY). DALY combines YLL and YLD and is a measure of healthy life years lost and as such is a more powerful metric to assess the burden and impact of CVD. One DALY equates to one year of healthy life lost and as such is a true measure of total disease burden [15].

MEASURING THE GLOBAL BURDEN OF CARDIOVASCULAR DISEASE

The generation of global estimates of disease burden is integral for the planning, design, and subsequent evaluation of health policies and interventions. Furthermore, it allows appropriate allocation of funding within health care systems on the basis of disease and mortality burden [16]. Many epidemiological studies have been performed since the 1940s following the prospective studies on IHD in professional men in Minnesota [17]. These studies though large do not provide a global overview of the current CVD global disease burden.

Limitations of the Quality and Completeness of Data

The WHO global health estimate studies aim to present a global measure of disease burden and inform international public health policy and planning.

Cause-of-death statistics are presented annually to the WHO. However, the ICD-10 classification is not used by every member state, and only 60 countries meet the inclusion criteria to admit statistics to the WHO mortality database [14]. Unfortunately, these statistics are not all coded to the same standard or for the same duration [14]. To account for this some statistical modeling is applied to the data that did not meet the high standard of the inclusion criteria [14]. Specifically, if coding of CVD data is poor, then it is redistributed to other diagnoses [14]. Where year data is missing interpolation and extrapolation is used to model for the missing year [14].

Many WHO member states have national vital registration systems that allow the WHO to access real mortality data. The data that is recorded from complete vital registration systems is not always accurate, and in one recent meta-analysis at least one-third of death certificates were incorrect [18]. A study carried out in East Germany, where for a year the outcome of all postmortem results were compared to death certificates issued, identified that in 47% cases there was a disparity. Interestingly the study identified that disease of the cardiovascular system was more likely to be overdiagnosed [19]. In contrast, Ravakhah et al. demonstrated that 48% of autopsy-proven acute myocardial infarctions were misdiagnosed as something else [20]. Furthermore, up to 30% of cardiovascular codes may need to be adjusted in certain countries who have been listed as IHD as a cause of death [16]. Thus, even the highest standards of coding systems that account for a large proportion of global statistics are partially inaccurate [16].

The United Nations proclaims the right of every individual to have a name and nationality recorded [21]. Despite this fundamental principle, nearly 50 million of all newborn children are not registered [21]. Outside of North America and Europe, less than one-third of all countries have usable mortality statistics, and nearly half have no mortality values at all [21]. The presence of an actual death certificate is a rarity in the developing world, with many countries not having a functioning registration system [16]. There are 194 WHO member states, and of these only 23 countries have actual data on >90% of the population [16]. In Africa, 25 of the 46 WHO member states could not provide any mortality data [16]. Annually the WHO receives actual data on 18.6 million deaths, estimated to be one-third of the total number of deaths worldwide. However, the mortality data provided have some important flaws that impair quality of reporting [16]. These include, but are not limited to, religious, political, and socioeconomic factors. Additionally, many people die at home without medical attention, making categorization of mortality difficult [16]. There are also factors that make information recorded on death certificates problematic [16].

Many WHO member states are unable to complete a vital registration system, due to the cost and infrastructure involved in running and maintaining such a system. Alternative systems are in use to attempt to increase recorded mortality data where a complete vital registration system is unfeasible, to allow relatively accurate mortality predictions to be made. China and India use a sample vital registration system, which monitors a random representative subset of the population over a period of time [16]. The representative sample mortality rate is applied at the population level to provide a national estimate. In China the Disease Surveillance Point system is used, which represents 1% of the population [22]. In India a nationally representative household survey has been in place since 1998 and monitors the cause of death in a population of over 14 million [23]. Epidemiological modeling is performed for countries where little or no mortality data is available. When limited data (eg, registry) is available, this information is used to inform a prediction model [16]. The GBD 2010 utilized this method for generating burden-of-disease metrics. In situations where no previous data is available, other small sources like verbal autopsies, disease surveillance systems, and analyses from WHO technical programs are used [14,16].

The ability to generate accurate figures on the global burden of CVD is challenging, and the most that can be offered is a best estimate. Unquestionably, the figures for lower income countries are more flawed, however, the presence of reasonably robust prediction models do provide good estimates. Until the infrastructure is in place to offer complete vital registration systems, these remain the best figures available. As such, interpretation of all figures presented must be performed with the quality of source information in mind.

CHANGES IN TOTAL DISEASE BURDEN VERSUS CVD

There have been significant proportional changes in the burden of disease between 1990 and 2010. Fig. 4.1 illustrates the top causes of loss of DALY and the proportional changes in their global burden within the prescribed period [2]. The majority of NCD burden has reduced globally. When comparing the top causes of loss of DALY, the greatest single increase in burden was due to human immunodeficiency virus [24]. The top two causes of loss of DALYs were IHD and stroke [2,24]. The greatest increase in burden between 1990 and 2010 was for atrial fibrillation (AF) and peripheral vascular disease (PVD) [2].

GLOBAL CARDIOVASCULAR MORTALITY

NCDs account for over 50% of all worldwide mortality annually, and of this, CVD represents the largest proportion, an estimated 31.4% in 2012 [4]. This has increased from 2000, when the estimated global mortality attributable to CVD was 28.2%. Table 4.1 outlines the global mortality data for CVD, comparing rates between 2000 and 2012 [2].

In 2000, IHD was the highest cause of mortality at 5.9 million, accounting for 11.3% of all mortality [4]. This increased over the following 12 years to an estimated 7.3 million (13.2%) deaths in 2012 [4]. The second commonest cause of death was stroke, responsible for 6.6 million deaths in 2012, representing 11.9% of global mortality [4]. Table 4.2 demonstrates the top 10 causes of global mortality in 2012 and compares them to 2000 [4]. There are three CVD categories in the top 10 causes of global mortality in 2012, not including diabetes mellitus [4].

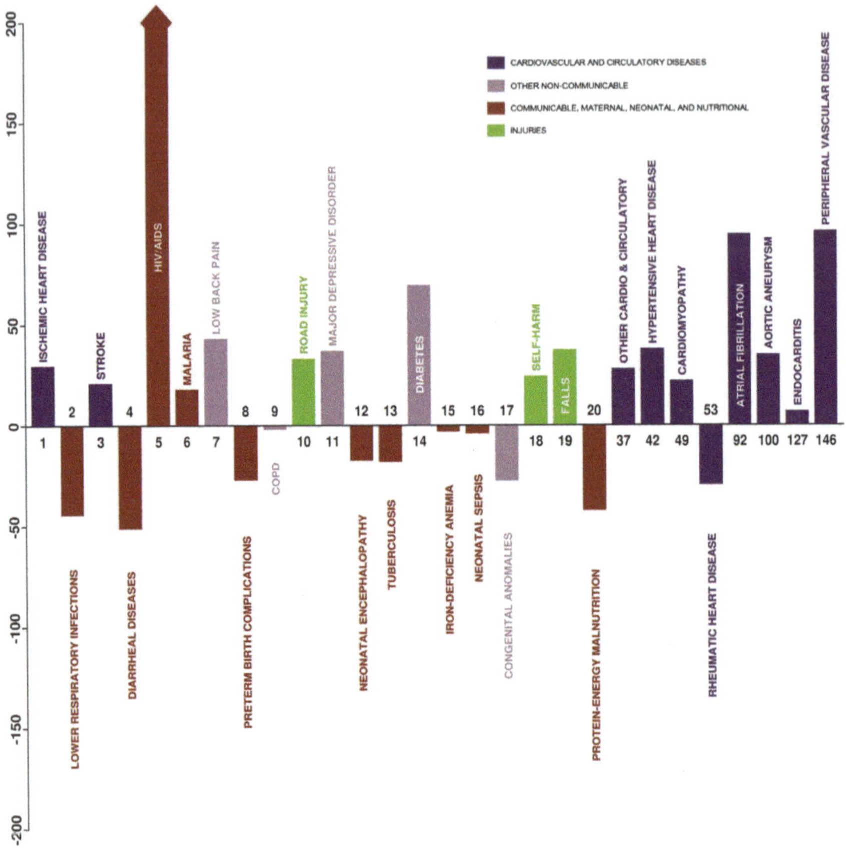

FIGURE 4.1 Cardiovascular disease in comparison with other leading causes of loss of disability-adjusted life years, global percentage change, 1990–2000 [2].

Global CVD Mortality Rates by WHO Region

Geographically the mortality burden for CVD varies widely. Fig. 4.2 demonstrates the WHO regions' mortality rate for 2000 and 2012. Overall, Europe carried the greatest mortality burden in 2012 due to CVD. There is a decrease from 2000, but CVD still accounts for nearly half of all mortalities in the WHO European Region. The mortality rate is significantly greater than the global rate calculated. Within Europe itself, there is considerable heterogeneity in the burden of CVD across regions, Eastern Europe being the driving force behind high CVD mortality rates [14]. The highest per capita CVD burden is attributed to Eastern Europe and Central Asia [2]. Explanation for some of these wide variations can be attributed to the prevalence of classic risk factors for CVD, like tobacco smoking, high blood pressure, and dietary risks [2].

The WHO regional groupings are based purely on geography and do not account for populations or socioeconomic factors. Proportionally, Europe as

TABLE 4.1 Global CVD Mortality Between 2000 and 2012

	2012		2000	
	Deaths (000s)	**% Total**	**Deaths (000s)**	**% Total**
All causes	55,859	100	52,806	100
Communicable, perinatal, maternal, and nutritional	12,822	23.0	16,310	30.9
Noncommunicable disease	37,892	67.8	31,470	59.6
Cardiovascular disease	17,519.3	31.4	14,881.5	28.2
Rheumatic heart disease	337.3	0.6	371.9	0.7
Hypertensive heart disease	1,141.2	2.0	849.0	1.6
Ischemic heart disease	7,356.1	13.2	5,974.1	11.3
Stroke	6,670.9	11.9	5,661.5	10.7
Cardiomyopathy, myocarditis, endocarditis	472.3	0.8	401.5	0.8
Congenital heart anomalies	230	0.4	234	0.4
Other circulatory diseases	1,541.5	2.8	1,623.5	3.1

World Health Organization. Comparison of specific CVD conditions. World Health Organization, Global health estimates 2014 summary tables: Deaths by cause, age and sex, by WHO region, 2000–2012 (Internet). 2014. Available from: http://www.who.int/healthinfo/global_burden_disease/en/; (accessed 23.03.15).

a region had the highest mortality in 2000 and 2012 attributable to CVD. The regional classification for Europe by the WHO includes Western Europe and Russia, both of which have completely differing prevalence of CVD mortality. In 2012 Russia had an estimated 1.3 million CVD mortalities, which accounted for 59.6% of all deaths [25], compared to the United Kingdom, where there were an estimated 169,855 CVD deaths, proportionally accounting for 30.5% of all mortalities in 2012 [25]. Both these countries are classified as Europe by region in the WHO classification. However, the cardiovascular mortality burden differs greatly, due to multiple factors including socioeconomic status, disease risk factor exposure, and population specifics. Analysis of the Americas demonstrated that CVD mortality for the region was 33.7% in 2000, with particularly high rates amongst women [26]. Over the last 12 years the Americas have seen a 20% reduction in mortality, which has mainly been driven by the high-income (HI) countries in the region. A further study in Brazil observed CVD death rates were falling, but slower than wealthier countries [27]. Such large variations amongst countries in the same region are observed and are likely to be accountable for, at least in part, by access to health care and local preventative policies.

TABLE 4.2 Top 10 Global Causes of Mortality in 2000 and 2012

Rank	Cause	Deaths (000s)	% Deaths	Deaths/100,000 Population
2012				
0	All causes	55,858.72	100.00	789.47
1	Ischemic heart disease	7,356.06	13.17	103.97
2	Stroke	6,670.93	11.94	94.28
3	Chronic obstructive pulmonary disease	3,104.33	5.56	43.87
4	Lower respiratory infections	3,051.99	5.46	43.13
5	Trachea, bronchus, lung cancers	1,599.56	2.86	22.61
6	HIV/AIDS	1,533.76	2.75	21.68
7	Diarrheal diseases	1,497.72	2.68	21.17
8	Diabetes mellitus	1,497.37	2.68	21.16
9	Road injury	1,254.53	2.25	17.73
10	Hypertensive heart disease	1,141.21	2.04	16.13
2000				
0	All causes	52,806.02	100.00	862.30
1	Ischemic heart disease	5,974.06	11.31	97.55
2	Stroke	5,661.55	10.72	92.45
3	Lower respiratory infections	3,490.57	6.61	57.00
4	Chronic obstructive pulmonary disease	3,059.32	5.79	49.96
5	Diarrheal diseases	2,171.25	4.11	35.46
6	HIV/AIDS	1,678.22	3.18	27.40
7	Tuberculosis	1,343.08	2.54	21.93
8	Preterm birth complications	1,316.10	2.49	21.49
9	Trachea, bronchus, and lung cancers	1,164.43	2.21	19.01
10	Diabetes mellitus	1,045.76	1.98	17.08

World Health Organization. Comparison of rates between regions 2014. Comparison of specific CVD conditions World Health Organization, Global health estimates 2014 summary tables: Deaths by cause, age and sex, by WHO region, 2000–2012 (Internet); 2014. Available from: http://www.who.int/healthinfo/global_burden_disease/en/; (accessed 23.03.15).

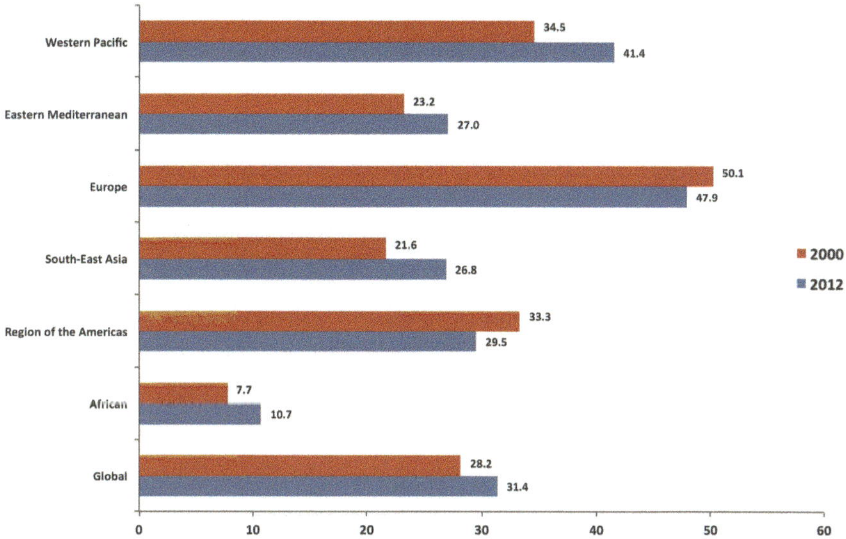

FIGURE 4.2 Cardiovascular Disease Mortality Burden by World Health Organization Region in 2000 and 2012. *World Health Organization. Comparison of rates between regions 2014. World Health Organization, Global health estimates 2014 summary tables: Deaths by cause, age and sex, by WHO region, 2000–2012 (Internet); 2014. Available from:* http://www.who.int/healthinfo/global_burden_disease/en/; *(accessed 23.03.15).*

Global CVD Mortality Rates by WHO/World Bank Region

The health status of a society has historically been linked to its economic progression and social constructs. Geographical analysis alone of population CVD does not account for certain influencing factors such as the disparity of wealth between nations and therefore the incongruity of available health resources. The WHO performs a regular socioeconomic analysis of disease burden using the World Bank regional grouping classification [11,28]. The assignment of countries to high-, middle-, and low-income designations uses the World Bank analytical income of economies for the fiscal year 2014 based on the 2012 Atlas of gross national income per capita estimates [11]. Fig. 4.3 compares mortality rates for NCD and CVD for HI- and LMI-designated countries (by region) for 2000 and 2012.

Age and Gender Global CVD Mortality Rates

As expected, global CVD mortality burden increases with age for both genders. Fig. 4.4A demonstrates mortality rates for each age group divided by gender for 2012. Overall, males carry the higher burden as they get older, until females pass menopause, when CVD mortality burden accelerates more rapidly for females. There is significant variation of mortality rates within each gender based on age and region. However, what is consistent across all regions and genders is

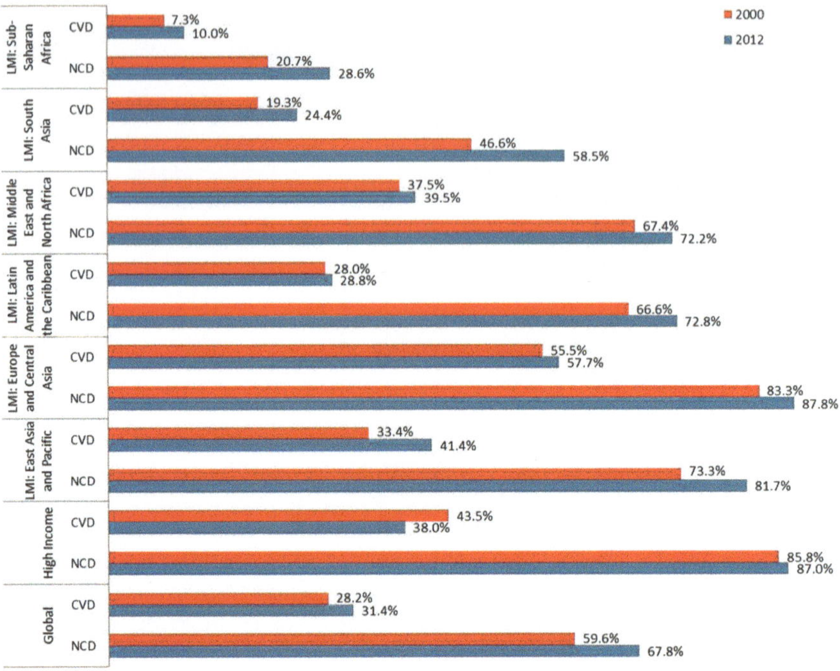

FIGURE 4.3 Mortality rates for noncommunicable disease and cardiovascular disease per World Bank Economic region, comparing 2000 to 2012 [11].

that there is a rapid significant increase in CVD over the age of 60. Notably the rate of mortality for CVD for the African region is the lowest for both males and females in older age groups and reflects an alternative health burden that is faced. The rate of acceleration in the mortality rate starts earlier in the South Asian region for males (>30 years) compared to others. This acceleration in rate of CVD mortality is not seen in South Asian females. This is likely related to the much higher rate of tobacco smoking in South Asian males compared to females [1].

GLOBAL CARDIOVASCULAR DALY

CVD accounted for an estimated 393.8 million (14.4%) DALYs in 2012, which increased from 352.8 million (12.3%) in 2000 [8]. DALY is a measure of total health impact from a particular condition and allows comparisons to be made globally, regionally, and nationally. Table 4.3 compares lost DALYs between 2000 and 2012 for major categories of conditions, CVD as a whole, and as specific conditions [8]. The WHO global health estimates data suggest that overall globally total lost DALYs have actually marginally decreased, the contribution of noncommunicable disease having increased, especially CVD. The specific conditions that have increased in terms of their global burden are IHD and stroke [8].

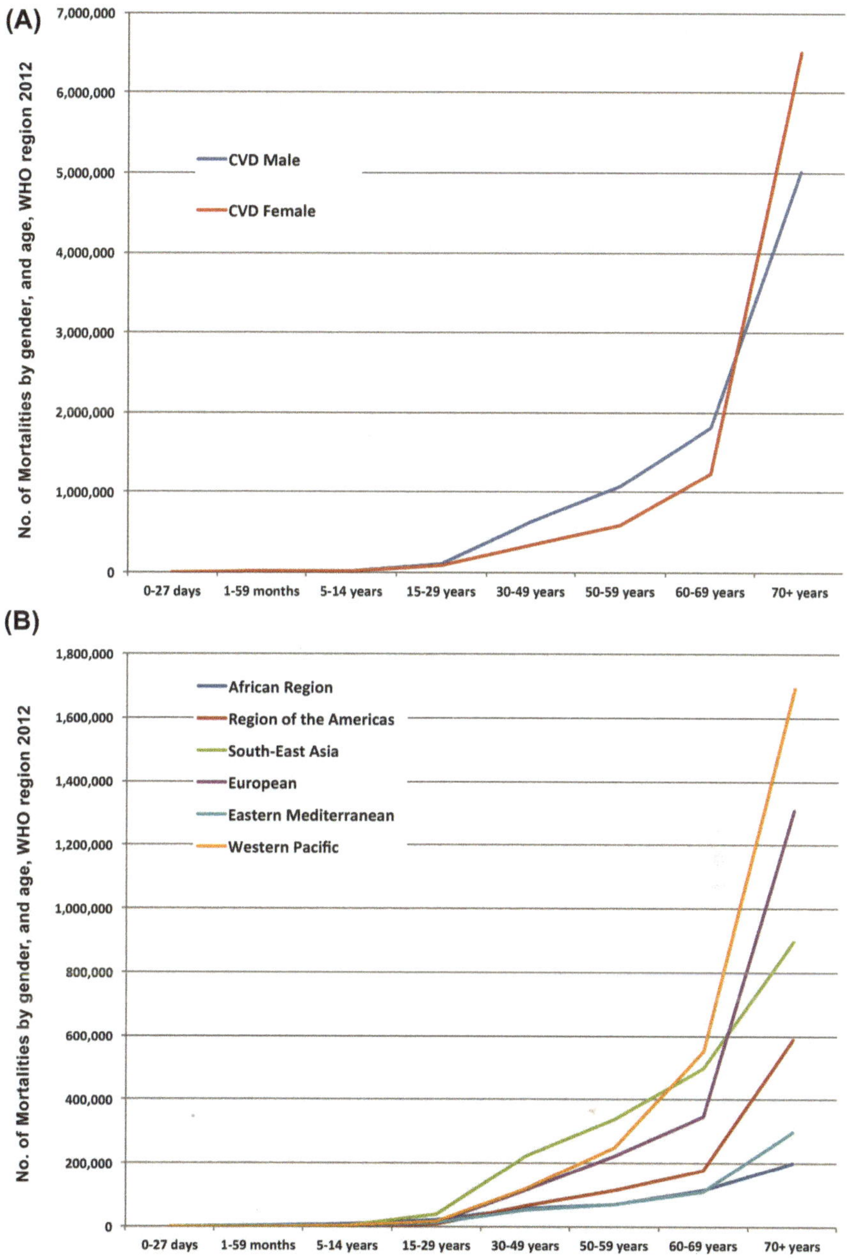

FIGURE 4.4 Global cardiovascular disease mortality burden by sex and gender. (A) Both genders, (B) Males per World Health Organization (WHO) Regions Source (C) Females per WHO Regions: WHO. Comparison of rates between regions 2014 [4].

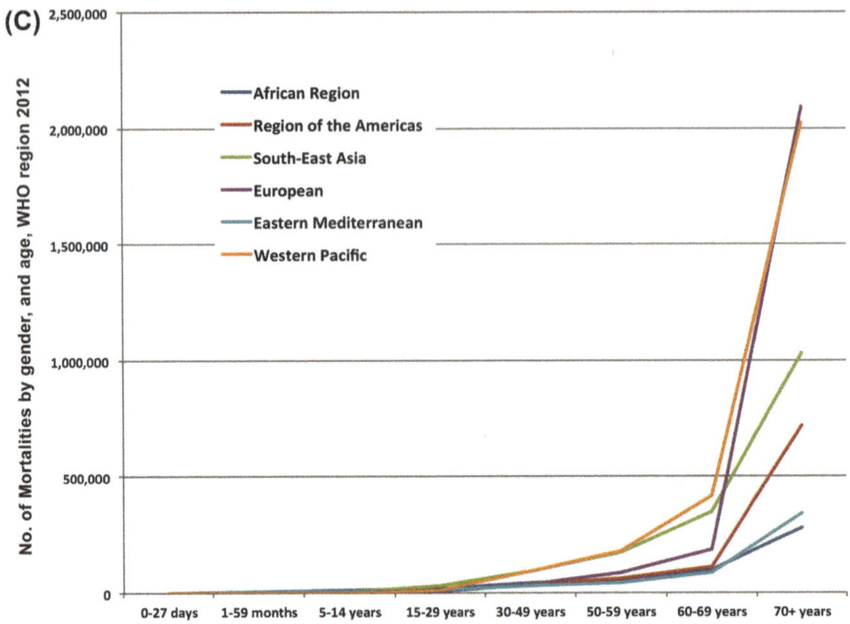

FIGURE 4.4 Cont'd

The top 10 highest DALYs in 2012 demonstrated a different set of diagnoses to the mortality figures. Utilizing the nonfatal contribution to global health there is the presence of conditions that globally have a significant impact but do not necessarily produce a hard end point like mortality. A death is the most serious of implications for a particular condition, but it does not inform us the extent of disability, temporal impact, or total effect. Table 4.4 demonstrates the top 10 highest DALYs attributable to a single cause. The top 10 causes of lost DALYs accounts for chronic medical conditions and for mental health status. The top cause of lost DALYs was lower respiratory chest infections in 2000 [8]. However, IHD and stroke move up the rankings in 2012 to occupy the first and third position, respectively [8]. The change in rankings demonstrates that CVD has become the largest fatal and nonfatal health burden worldwide [8].

Global CVD Lost DALYs Within WHO Regions

The fatal and nonfatal CVD burden varies regionally and between countries. Fig. 4.5 demonstrates the number of DALYs lost per 100,000 population in 2010 per country. The largest burden falls on Eastern Europe and Russia [2]. Areas in central Africa have some of the lowest rates of lost DALYs attributable to CVD [2]. These rates mirror the mortality rates for 2012 within the WHO African Region, which are the lowest for CVD.

TABLE 4.3 Global Cardiovascular Disease Disability-Adjusted Life Year Rated Between 2000 and 2012

	2012		2000	
	DALYs	Percentage (%)	DALYs	Percentage (%)
All causes	2,743,857,491	100.0	2,872,910,085	100.0
Communicable, maternal, perinatal, and nutritional conditions	925,727,921	33.7	1,243,421,591	43.3
Noncommunicable disease	1,512,577,728	55.1	1,315,770,489	45.8
Cardiovascular disease	393,804,862	14.4	352,854,272	12.3
Rheumatic heart disease	11,953,850	0.4	14,312,769	0.5
Hypertensive heart disease	23,408,695	0.9	19,246,058	0.7
Ischemic heart disease	165,717,210	6.0	142,225,562	5.0
Stroke	141,348,082	5.2	125,127,132	4.4
Cardiomyopathy, myocarditis, and endocarditis	14,603,113	0.5	13,537,411	0.5
Congenital heart anomalies	19,784,491	1.3	20,291,404	0.7
Other circulatory diseases	36,773,908	1.3	38,405,340	1.3

World Health Organization. Comparison of specific rates for each CVD condition. World Health Organization, Global health estimates 2014 summary tables: DALY by cause, age and sex, by WHO region, 2000–2012 (Internet); 2014. Available from: http://www.who.int/healthinfo/global_burden_disease/en/; (accessed 18.04.15).

TABLE 4.4 **Top 10 Global Causes of Disability-Adjusted Life Years (DALYs) in 2000 and 2012**

Rank	Cause	% DALYs	DALYs/100,000 Population
2012			
0	All causes	100	38,779.9
1	Ischemic heart disease	6.0	2,342.1
2	Lower respiratory infections	5.4	2,075.7
3	Stroke	5.2	1,997.7
4	Preterm birth complications	3.9	1,515.2
5	Diarrheal diseases	3.6	1,409.5
6	Chronic obstructive pulmonary disease	3.4	1,305.6
7	HIV/AIDS	3.4	1,299.0
8	Road injury	2.9	1,112.6
9	Unipolar depressive disorders	2.8	1,081.2
10	Birth asphyxia and birth trauma	2.7	1,054.3
2000			
0	All causes	100	46,913.5
1	Lower respiratory infections	7.3	3,402.0
2	Diarrheal diseases	5.6	2,624.5
3	Ischemic heart disease	5.0	2,322.5
4	Stroke	4.4	2,043.3
5	Preterm birth complications	4.3	2,015.9
6	Birth asphyxia and birth trauma	3.6	1,666.1
7	HIV/AIDS	3.5	1,659.8
8	Chronic obstructive pulmonary disease	3.1	1,460.2
9	Malaria	2.7	1,277.7
10	Road injury	2.4	1,129.0

World Health Organization. Comparison of rates between regions 2014. World Health Organization, Global health estimates 2014 summary tables: DALY by cause, age and sex, by WHO region, 2000–2012 (Internet); 2014. Available from: http://www.who.int/healthinfo/global_burden_disease/en/; (accessed 18.04.15).

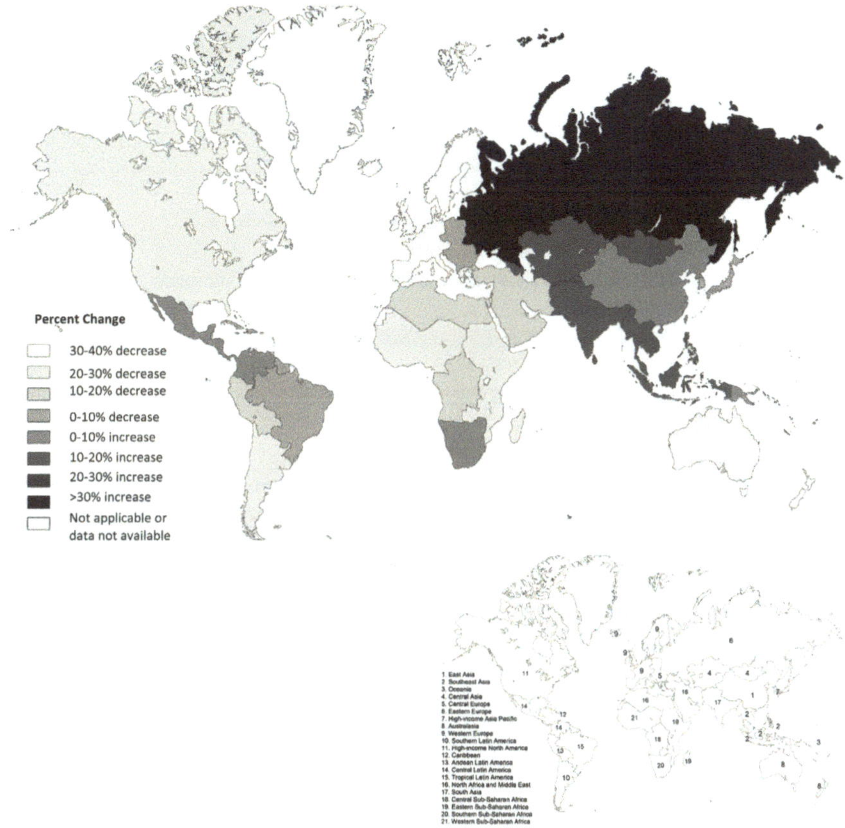

FIGURE 4.5 Age-standardized cardiovascular disease disability-adjusted life years lost per 100,000 for 2010 [2].

Global CVD Lost DALYs by WHO/World Bank Region

Factoring in the economic status of the WHO member states, the fatal and nonfatal NCD and CVD burden varies between different World Bank regions. Fig. 4.6 compares the rates of DALYs for NCD and CVD of the total per region in both 2000 and 2012. In all the LMI regions the number of DALYs attributable to NCD and CVD increase from 2000 to 2012. Southeast Asia has the greatest increase in proportion of DALYs gained for NCD. HI countries have over 80% of their DALYs attributed to NCD in both 2000 and 2012 [28]. DALYs attributable to CVD actually decrease from 2000 to 2012. This pattern mimics that of NCD and CVD mortality rates as demonstrated in Fig. 4.3.

Age and Gender Global CVD Loss of DALYs

Loss of DALYs provides a different measure of the global impact of CVD and the trends vary significantly. Fig. 4.7A outlines the number of DALYs globally

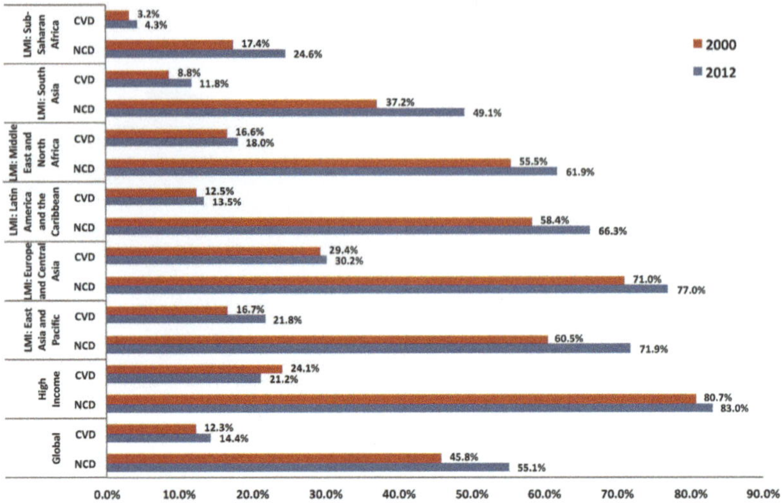

FIGURE 4.6 Percentage noncommunicable disease and cardiovascular disease of total disability-adjusted life years 2000 and 2012 by World Banking Region [28].

attributable by gender and age. Overall, males have 216.1 million (14.5%) and females have 177.7 million (14.1%) DALYs attributed to CVD [8]. The predominant cause behind the high number and proportion of DALYs in males is IHD. In contrast, DALYs amongst females are driven by a combination stroke and IHD. The leading cause of DALYs in females is stroke, rather than IHD [8]. Fig. 4.7B and C demonstrated the variation for males and females in different regions by age. Overall, both genders behave in a similar way; all regions lose more DALYs with progressive age. The Western Pacific region for both genders has the greatest loss of DALYs with age [8]. European females have a more rapid loss of DALYs in later years, in contrast to males who have a greater loss of DALYs in earlier and middle life [8]. The Southeast Asia males have a more rapid loss of DALYs in earlier life than females [8].

RISK FACTORS FOR CVD GLOBALLY

The majority of CVD burden can be attributed to a relatively small number of risk factors. These include, but are not limited to, diet, exercise, tobacco smoking, hypertension, dyslipidemia, and alcohol consumption [26]. The regional variation in the presence of particular lifestyle risk factors varies between countries and cultures influencing CVD burden.

Hypertension

Worldwide, hypertension contributes to 7.5 million deaths (12.8%) and 5.7 million DALYs (3.7%) [1,29]. High blood pressure is the leading cause of

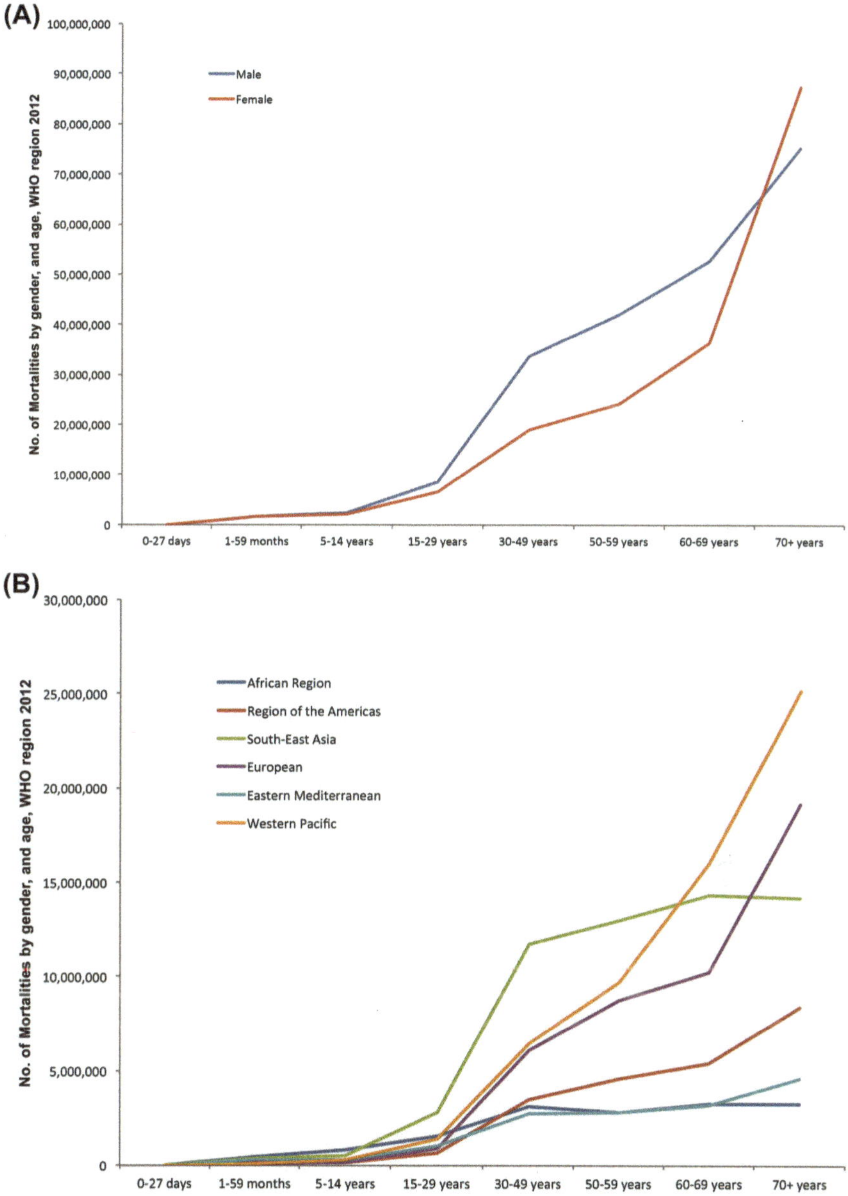

FIGURE 4.7 Total disability-adjusted life years for global cardiovascular disease by sex and age in 2012 (A) Genders globally, (B) Men by age and World Health Organization (WHO) Region, (C) Females by age and WHO region [8].

(C)
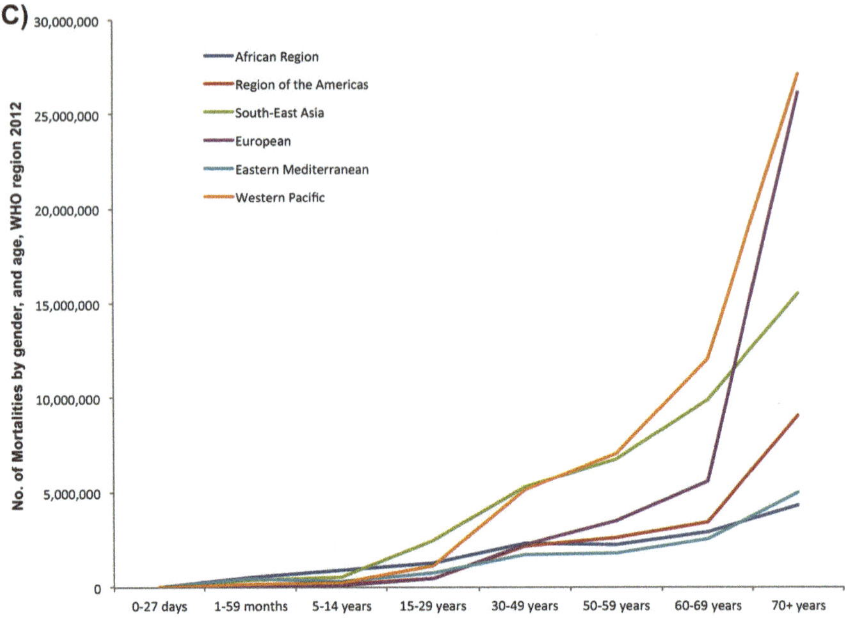

FIGURE 4.7 Cont'd

attributable risk factors to death [30]. Hypertension is a well-established global risk factor for both IHD [31] and cerebrovascular disease [1]; in addition it is known that incremental increases in blood pressure increase the risk of CVD [31]. Furthermore, if untreated, hypertension can increase the risk of developing heart failure, renal disease, peripheral vascular disease, and retinal vessel injury [1,31,32]. The number of people with uncontrolled hypertension increased from 600 million in 1980 to almost 1 billion in 2008 [29].

Global hypertension prevalence varied significantly with the highest rate in the WHO African region at 46% and the lowest rate in the WHO Americas region at 35% [29]. Western Europe had a higher prevalence of hypertension in males than females [1]. The opposite is true in Southeast Asia, with females having a higher prevalence of hypertension than males [1].

Diabetes Mellitus

Diabetes was estimated to affect 171 million (2.8%) of the population in 2000 and is projected to more than double by 2030 to 366 million (4.4%) [33]. CVD accounts for approximately 60% of all deaths of people with diabetes [1]. Worldwide, diabetes accounted for nearly 1.5 million deaths in 2012, which increased from just over 1 million in [11]. Incredibly, nearly one-third of total worldwide mortality due to diabetes occurs in South Asia. The global impact of diabetes was over 59 million DALYs, which is 2.2% of the total DALYs in

2012 [4]. The risk of cardiovascular events with type 1 or type 2 diabetes is two to three times higher than in the unaffected population [34]. The risk is higher for women than men [1]. In older age groups diabetes can increase the risk of stroke by more than twofold [35]. Furthermore, the mere presence of diabetes increases the risk of a poor outcome following a cardiovascular event [1] and has consistently been shown to be an independent predictor of both mortality and morbidity in heart failure [36–40].

Worldwide the number of people affected by diabetes is difficult to calculate, especially for developing countries [33]. An important note in the analysis of diabetes prevalence is that many data sources do not separate type 1 and 2 diabetes, therefore many estimates are based on diabetes as a whole [33]. Worldwide the numbers of people with diabetes is expected to double between 2000 and 2030, with the biggest increases seen in the Middle East, Sub-Saharan Africa, and India [33]. Increasing age leads to significantly increased risks of developing diabetes; it is estimated that the number of people with diabetes >64 years of age will be >82 million in developing countries and >48 million in developed countries by 2030 [33]. Table 4.5 demonstrates the numbers of diabetics in 2000 and projected estimates for 2030 for the 10 top-ranking countries.

TABLE 4.5 List of Countries With the Highest Numbers of Estimated Cases of Diabetes for 2000 and Projected 2030 [33]

	2000		2030	
Ranking	Country	People With Diabetes (Millions)	Country	People With Diabetes (Millions)
1	India	31.7	India	79.4
2	China	20.8	China	42.3
3	United States	17.7	United States	30.3
4	Indonesia	8.4	Indonesia	21.3
5	Japan	6.8	Pakistan	13.9
6	Pakistan	5.2	Brazil	11.1
7	Russia	4.6	Bangladesh	11.1
8	Brazil	4.6	Japan	8.9
9	Italy	4.3	Philippines	7.8
10	Bangladesh	3.2	Egypt	6.7

Tobacco

Smoking tobacco poses one of the greatest risks for development CVD [29]. Worldwide there are 1 billion smokers and it is estimated that 6 trillion cigarettes are consumed annually [29,41]. Globally, 9% of all NCD mortalities are attributable to tobacco [29]. The highest prevalence of smoking is in the WHO European region and is estimated to be 31%, with the lowest estimated to be 10% in the WHO African region [42]. Tobacco accounts for 12% and 7% of total male and female deaths, respectively [30]. The total CVD attributable to smoking tobacco is estimated to be nearly 10% [30].

Globally there are wide variations in the prevalence of smoking amongst the population that adds to the CVD burden for the region. Fig. 4.8 identifies the worldwide prevalence of daily tobacco smoking in 2011 by gender [1]. Overall the prevalence of smoking is the same in Eastern Europe and Russia for both

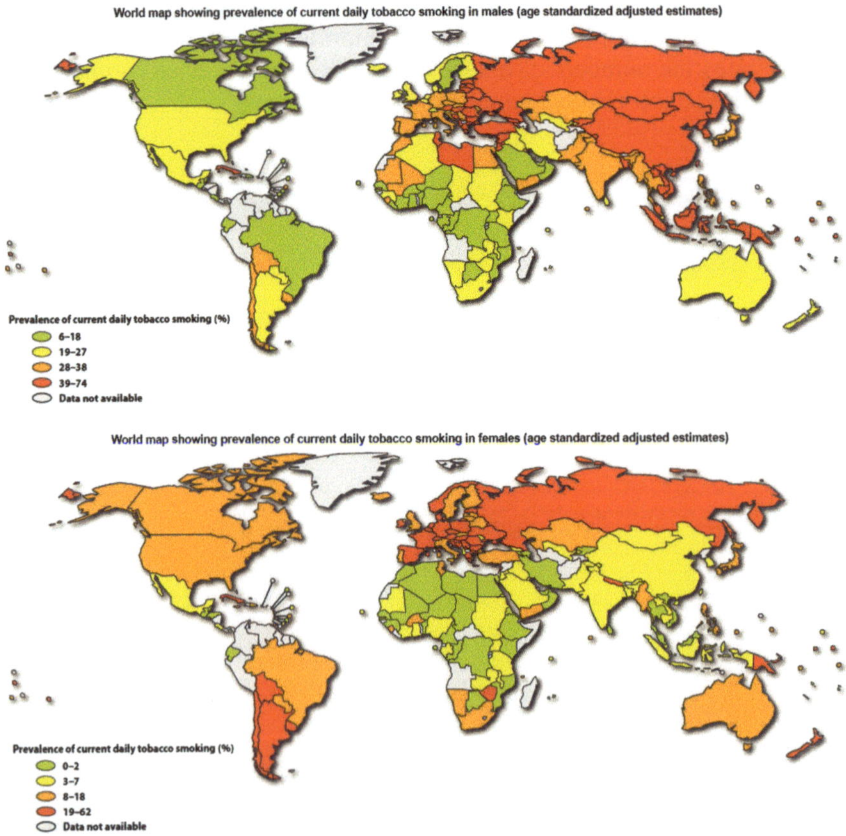

FIGURE 4.8 World map demonstrating the prevalence of current daily tobacco smoking by gender in 2011 [1].

genders [1]. A higher proportion of males smoke in Southeast Asia than females [1], whereas in Europe the trend is reversed [1].

Smoking rates in the United States have been declining since the 1960s, when the warnings about health concerns with tobacco were first published. In 1965, 41.7% of adults smoked, compared to 19.3% in 2010 [3]. In contrast, in more populous areas of the world like Eastern Europe and Southeast Asia, smoking is the third largest risk factor for CVD [2].

Dyslipidemia

Hypercholesterolemia increases the risk of IHD and stroke [35]. Worldwide, raised cholesterol is associated with one-third of all IHD deaths [1] and dyslipidemia is estimated to be the cause of 2.6 million (4.5%) deaths and 29.7 million (2%) lost DALYs [30]. In 2008, the global rate of raised total cholesterol was 39% (male 37% and females 40%) [1]. Over 50% of the population of HI countries have an elevated total cholesterol. There has been no real change in mean total cholesterol between 1980 and 2008, with less than 0.1 mmol/L fall per decade, for both males and females [1]. Although mean total cholesterol is highest in HI countries, they have seen the greatest mean drop from 5.62 mmol/L to 5.19 mmol/L. The change has been much lower in LMI countries [3]. The prevalence is highest in the WHO European region at 54% (both genders), followed by the Americas at 48% (both genders). The lowest prevalence was Africa at 23% and Southeast Asia at 30% (for both genders) [29].

Physical Activity

Insufficient activity is defined as less than 30 min of moderate activity per week. Lack of physical activity is a leading cause of CVD, especially IHD [3,43]. The impact of physical inactivity on the global health burden is 3.2 million deaths and 32.1 million DALYs [30]. Individuals who have insufficient physical activity have a 20–30% increased risk of all mortality [1]. Participation in physical activity reduced the risk of IHD and stroke in a dose-dependent fashion [44].

The WHO region with the greatest prevalence of inactivity is the Region of the Americas [1]. The prevalence of inactivity in HI countries is more than double that of LMI countries in both men and women. The HI countries have an estimated 41% and 48% of men and women, respectively, who are inactive. In contrast, 18% of men and 21% of women are inactive in lower income countries [29]. Individual countries do not have a significant variation in inactivity rates between the genders [1].

Alcohol Consumption

Alcohol consumption is associated with the development of CVD [43]. The dangerous consumption of alcohol was responsible for 2.5 million deaths in

2004; over half of these deaths were due to CVD, liver cirrhosis, and cancer [1,30]. In all regions of the world it contributes to the prevalence of CVD, with a more significant contribution in LMI countries [1]. Eastern Europe has high alcohol consumption as a population and it is the fifth largest contributor to CVD [24]. Interestingly, in other regions alcohol does not rate as highly as a contributor to CVD burden [2].

The evolution of the relationship between CVD and alcohol consumption has changed in the last 30 years from one where it was traditionally thought to be cardioprotective to one where it can contribute to the CVD disease burden. This view was immortalized by the 1996 WHO/World Bank report on the global burden of disease that "alcohol is cardio-protective at all levels of consumption" [45]. Studies in the same period from Russia and Poland demonstrated high cardiovascular mortality rates in young men who had low traditional risk factor exposure but were heavy drinkers [46,47]. Interestingly, France during this period had one of the lowest mortality rates but reported alcohol consumption similar to Russia [48].

A large meta-analysis performed to reconcile the conflict between the cardioprotective role of alcohol outlined in the West and the association between consumption and CVD in Eastern Europe demonstrated that problematic drinking was the key risk factor [48]. The analysis demonstrated that there was a strong association between alcohol consumption, arrhythmias, and sudden cardiac death [48]. A case–control study from Moscow demonstrated a higher rate of sudden cardiac deaths at the weekends, coinciding with higher rates of alcohol consumption [49]. These findings have since been replicated in a more recent Scottish study [50]. Patterns of alcohol consumption are causative for CVD and in particular sudden cardiac death. Traditionally, Eastern Europe and Russia have drink in a heavy pattern. This will partly explain the greater significance of alcohol consumption in Eastern Europe.

EPIDEMIOLOGICAL TRANSITION

Progression of society through industrialization reflects in our changes in health status. Traditionally, populations tended to suffer from infectious disease and nutritional deficiencies. Improvements and developments in society have led to a switch in health status to NCD related to degenerative conditions [51]. The literature describes this process as the "epidemiological shift" [51]. Different countries, or even areas within the same country, can be at varying stages in this transition [51]. This transition can occur between distinct diseases or within the same disease category [51].

Countries in the earliest stages of development are expected to have a higher burden of rheumatic heart disease. Sub-Saharan Africa has one of the highest percentages of mortality related to rheumatic heart disease globally. In 2012 rheumatic heart disease accounted for 3.5% of CVD mortality in Sub-Saharan Africa compared to 1.9% globally [11]. Proceeding along the epidemiological transition, as the infection rates decrease and nutrition improves, the burden of hypertension prevalence increases [51]. This leads to increased rates of hemorrhagic stroke and hypertensive heart disease [51]. Life expectancy increases and smoking tobacco, high-fat diets, and sedentary lifestyle predominate [51]. This reflects the global impact of atherosclerotic disease and identifies why IHD and strokes are the commonest cause of death globally [51]. The transition progression demonstrates the increasing extent of NCD globally, especially in HI countries. Finally, as a consequence of better-developed health care systems, HI countries are able to diagnose and treat IHD and stroke, or their causal risk factors and thus delay negative outcomes [51]. The regions that are in this phase are North America, Western Europe, and Australia [51].

AGING POPULATION

Globally, individuals are living longer with an increasing life expectancy. In 1990 the global life expectancy was 64 years, which has increased in 2013 to 71 [52]. Between different regions there is wide variation in the average global life expectancy [52]. For 2013 the Americas have the highest average life expectancy at 77 years, which is in contrast to the lowest average life expectancy of 58 for Africa [52]. There are also differences between the genders with a global average life expectancy of 74 and 78 for males and females, respectively [52].

The increasing age of the population not only reflects the increasing global epidemiological transition but also the different health burdens. One particular health burden that increases is CVD, which is reflected in populations with higher life expectancy [2]. Figs. 4.4A–C and 4.7A–C demonstrate the impact of the aging population on the CVD risk; this particular nonmodifiable risk factor provides a higher burden on the better economically developed countries. In 2012 the CVD mortality for HI countries was 38.0%, which is in direct contrast with Sub-Saharan Africa at 10.0% [11]. Comparing a specific CVD condition, IHD, the difference is even greater. IHD is the commonest cause of death globally, and accounts for 17.5% of all HI countries' mortality, but in Sub-Saharan Africa it only accounts for 3.1% [11].

Eastern and Central Europe have a large aging population with over 10% aged >65 years [2]. The CVD mortality burden reflects this specific aging population. The overall CVD mortality for Eastern Europe and Central Asia regions in 2012 was 57.6%, higher than the HI countries [11]. Interestingly, the commonest causes of death in this region were IHD and stroke with a rate of 32.9% and 16.6%, respectively [11]. The worldwide population will continue to increase in age, and the proportion of those above the age of 65 years will also increase. In coming decades, regions like Latin America and the Caribbean are predicted to have a significant increase in population above the age of 65 years [2], which will result in an increasing CVD burden in the future.

ETHNICITY

The ethnic constructs of populations influence the CVD burden worldwide and are related to biological, cultural, health care, and social issues [53]. Several epidemiological studies have examined the variation within different patient populations to examine the impact globally of ethnicity. The INTERHEART study was a global case–control study of risk factors for acute myocardial infarction [43]. The study followed over 29,000 patients across 52 countries worldwide and examined for associations between a wide range of risk factors and acute myocardial infarction within different ethnic and geographical populations [43]. Nine important risk factors (apolipoprotein B, apolipoprotein A-I, current smoking, diabetes, hypertension, abdominal obesity, psychosocial index, lack of exercise, and excessive alcohol consumption) were associated with acute myocardial infarction, in all ethnic groups and populations [43].

European

European populations have diverse origins (Nordic, Western, Southern, and Eastern Europe) and have large differences in age-standardized mortality rates [54]. To exemplify this point, in both genders the cardiovascular mortality is sixfold higher in Russian than in France [54]. Eastern Europe has a higher rate of fatal and nonfatal CVD burden, which is in contrast to other European populations [54]. CVD mortality accounted for 58.5% and 30.9% of total populations' deaths in Romania and Spain, respectively [12]. The contrasts persist when comparing genders for both countries. Male CVD proportion of total mortality is 52.0% in Romania and 27.4% in Spain [12]. The contribution of CVD mortality on total deaths for females in Romania is 66.1% compared to Spain, which is 34.8% [12].

The risk profile for CVD in European populations is primarily attributable to the main lifestyle risk factors. The prevalence of risk factors across European populations varies widely [54]. The higher rates of CVD burden in Eastern Europe are linked to higher rates of smoking, alcohol consumption, diets high in saturated fats, and poor social circumstances [54]. A fascinating finding from CVD epidemiological studies is that Southern European populations remain relatively protected from IHD. Why is this? One hypothesis is the high consumption of monounsaturated fats (like olive oil), which may have a protective effect, resulting in lower rates of IHD in countries such as Italy. Another theory is that the high alcohol consumption, specifically wine, is related to lower incidence of IHD in France [55].

Changes in CVD prevalence have altered through the last decades [56]. Many Western European countries have demonstrated a reduction in CVD, which has been attributed to better primary and secondary prevention [56]. Moreover, improvement in medical and surgical care of patients with IHD has improved survival [1,56]. A prime example of this change is in Finland where IHD mortality was particularly high in the 1970s. Between 1972 and 1992 there was a 55% reduction in mortality due to IHD and stroke [54]. The driver behind this reduction was a lowering of serum cholesterol, lowering of diastolic blood pressure, and a reduction in tobacco smoking [57].

Chinese

The mortality rates in China from CVD have been increasing in recent decades [54]. Western Europe demonstrates 50% lower CVD mortality rates than China [54]. The Monitoring Trends and Determinants in Cardiovascular Disease (MONICA) study was a major international observational study of over 15 million patients including European and Asian nations [58]. MONICA identified that hemorrhagic stroke rates were two to three times higher in Chinese compared to white Caucasian populations [58]. Although serum cholesterol is typically lower in the Chinese population, it still directly correlates with IHD mortality [59]. Genetic factors may be at play, predisposing the Chinese to CVD even at lower levels of cholesterol. Smoking tobacco is particularly high amongst Chinese males and is a major contributing factor to the high CVD burden [54]. Migration patterns have been shown to impact the CVD burden on different geographical Chinese populations. An observational study in Ontario, Canada, demonstrated that the Chinese diabetic population had a lower CVD event rate than the European population [60].

South Asian

South Asian refers to people whom originate from India, Sri Lanka, Bangladesh, Nepal, and Pakistan [54]. This population group represents 20% of the global population and are a heterogeneous group with differences in diet, culture, and lifestyle [61]. CVD mortality in this region is challenging to calculate, specifically for India, because of inconsistency of mortality record keeping [54]. WHO/World Bank epidemiological modeling estimates India to have the second highest CVD mortality rate worldwide, which was estimated to be just over 2.5 million in 2012 [12]. The CVD burden has been modeled to take account of the rapidly increasing population in India [54]. Migration of South Asian populations to the United Kingdom and the United States demonstrated a 1.5–4.0-fold increased risk of IHD mortality compared with the native populace [54]. Within the United Kingdom, the South Asian population carries a higher CVD risk, however, this risk has started to decline over the last two decades [53,54]. A cohort study compared admission rates to hospital of newly diagnosed heart failure patients for South Asians and white Caucasians in Leicestershire, UK, over a 10-year period [62]. South Asians were found to have a higher hospital admission rate and tended to be younger and male [62]. Outcome data, surprisingly, demonstrated that South Asians had a lower risk of death. In Ontario, Canada, the South Asian population with diabetes demonstrates lower mortality rates, however, nonfatal cardiovascular complications were equivalent to the population of European origin. The impact of CVD on Southeast Asian countries can be modeled on population members who reside in other regions and the impact of CVD on them specifically, given the local risk factor exposure. For example, it is predicted that India will have an increasing CVD burden, and this is evident from the most recent CVD mortality rate [54].

Despite risk factor exposure in the South Asian population within the United Kingdom and Canada being similar to the indigenous white populations, the rates of IHD are higher [63,64]. Amongst South Asians there are particularly strong associations with smoking, hypertension, and diabetes mellitus with IHD [65]. Metabolic syndrome is more prevalent among South Asians with higher rates of impaired glucose intolerance, obesity, elevated triglyceride levels, and type 2 diabetes mellitus [54,63]. Worldwide, South Asians carry high rates of type 2 diabetes; United Kingdom 10–19%, Trinidad 21%, Fiji 25%, and South Africa 22%. Yet India only has a reported rate of 2–3% in rural areas and 8% in urban areas [66]. The presence of diabetes mellitus in India is predicted to increase from 2000 to 2030 in the entire population by 40% [33]. The rural versus urban contrast in India is quite wide, with a previous prevalence survey over two decades demonstrating an increase in IHD rates of ninefold in urban areas versus twofold in rural areas [67]. The higher urban rates are associated with higher rates of hypercholesterolemia, prevalence of diabetes, obesity, and hypertension [54].

Hispanic

The Hispanic population includes Americans of Cuban, Mexican, and Puerto Rican descent, accounting for 16.3% of the US population [68]. CVD accounts for 28% of male and 34% female Hispanic mortality [54]. Relatively, the Hispanic population in the United States carries a higher incidence of acute myocardial infarction than non-Hispanic Americans [54]. This risk is significantly higher for the under-60 years of age population [54]. Hispanics within the United States have a higher prevalence of classic risk factors in both men and women, such as hypertension, smoking, elevated cholesterol, diabetes, and obesity [54].

The Hispanic Community Health Study enrolled >15,000 men and women to examine the disease burden and consequences of immigration and acculturation for the largest minority in the United States [69]. The study reported the greatest risk factor burden was recorded among Puerto Rican study participants, those with low socioeconomic status and those with the highest level of acculturation [70].

Blacks of African Origin

Sub-Saharan Africa is a region that varies completely from other regions and other populations of black people of African origin. Atherosclerosis only causes half of all CVD burden here, lower than all other regions [6]. The leading cause of death is stroke at 38.8% in 2010 [6]. The largest increases in disease prevalence between 1990 and 2010 was AF and PVD [6]. Sub-Saharan Africa has the lowest mortality rate due to CVD globally at 10.0% in 2012 [11]. Furthermore, CVD only accounts for 3.5% of total DALYs in the region. The epidemiological transition is in an earlier phase in this region, with infectious and inflammatory causes driving the disease burden of the region. This reflects in the lower life expectancy for the Africa region of 58 years for both genders [52]. The CVD burden has a different pattern and tends to occur at younger ages and causes mortality at a younger age [6]. PVD is common in Europe and North America above the age of 55, conversely, in Sub-Saharan Africa it tends to occur in the under 55 ages. Caution must be taken when analyzing information from Sub-Saharan Africa, as data depletion or absence is present and is based on estimates and epidemiological modeling [6]. Sub-Saharan Africa has the lowest number of years of life lost due to CVD worldwide, as adults are more likely to die of non-CVD causes at an earlier age [6]. South Africa has demonstrated rapid rates of urbanization with black members of the population rapidly migrating from rural communities to urban centers, leading to increased poverty, hypertension, and hypercholesterolemia [54]. Risk factor exposure in Sub-Saharan Africa is generally quite low, especially among the black population, compared to other groups residing in the region [54]. The prevalence of hypertension, hypercholesterolemia, and obesity are all low compared to other groups in Africa and beyond [71].

MIGRATION

The Nippon-Honolulu-San Francisco Study (Ni-Hon-San Study) demonstrated the impact of an ethnic group's migration on their IHD risk [72]. This cross-sectional study conducted in 1965 involved 20,000 Japanese men aged 45–69 years and compared those that migrated to the United States (Hawaii and California) with those that remained in Japan [72]. The study demonstrated that American Japanese men had higher cholesterol and higher IHD mortality than those still in Japan [72,73]. The study demonstrated the clear effects of lifestyle change as a result of acculturation on the impact of CVD risk [72,73].

URBANIZATION

Urbanization (movement of individuals from rural to urban environments) is increasing globally from 36.6% to 44.8% in 1970 and 1994, respectively. This is predicted to increase to 61% by 2025 [51]. With urbanization there is an increase in consumption of high-energy foods, decrease in exercise, and a loss of traditional social support mechanisms [51]. Moreover, the rural environment is evolving with increased mechanization of agriculture and increasing amounts of automobile transport, which is leading to decreased activity [51]. There are also external influences from media that are driving changes in the types of foods consumed [51]. These factors are likely to result in an increasing rural CVD burden.

SPECIFIC CVD CONDITIONS

Ischemic Heart Disease

IHD was the single greatest cause of fatal and nonfatal global disease burden in 2012 with an estimated 165 million lost DALYs, which accounts for 6.0% of the total [8]. The burden of IHD has increased from 2000, when the largest single cause of lost DALYs was lower respiratory chest infections accounting for 7.3% of the total [8]. Globally, there are large variations in total burden of IHD. The LMI Southeast Asia region accounted for the highest proportion of DALYs attributed to IHD in 2012 at 26% [28]. The LMI Africa and Americas region had the lowest proportion of DALYs attributable to IHD, both 6% [28]. Fig. 4.9A demonstrates the WHO/World Bank regions' contribution to the worldwide IHD DALY lost in 2012 [28].

The pattern of estimated variation in the IHD DALY rate changed between 2000 and 2012. The estimated total loss of DALYs for IHD worldwide in 2000 was over 142 million [28]. By 2012 there had been a significant increase in the number of lost DALYs for IHD worldwide [28]. Fig. 4.9B demonstrates WHO/World Bank regions' contribution to the worldwide IHD DALY lost in 2000 [28]. The majority of regions have seen small increases in numbers of DALYs lost to IHD from 2000 to 2012. The LMI Africa, Americas, and East Mediterranean regions display the smallest increase in loss of DALYs. There was an increase in numbers of lost DALYs, from over 35 to 42 million, amongst the LMI Southeast Asia group, though the proportion only increased by 1% [28]. The LMI European estimates are essentially static [28]. The largest shift in proportion is in the HI countries and the LMI Western Pacific, which accounts largely for the increase in total loss of DALYs worldwide between 2000 and 2012. The LMI Western Pacific region increased its estimated loss of DALYs in 2000–2012 from approximately 17 million up to over 35 million [28]. Contrasting this increase the total burden of IHD in high-income countries decreased in the same period from just over 46 million to over 39 million [28]. Worldwide

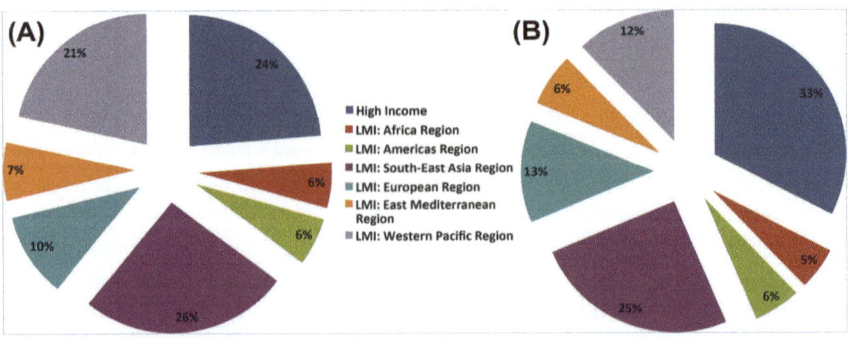

FIGURE 4.9 World Health Organization/World Bank Region contribution to total ischemic heart disease disability-adjusted life years. (A) in 2012 and (B) in 2000 [28].

the combined fatal and nonfatal burden of IHD increased, but certain regions in the last 12 years have changed their total burden. The reduction in HI countries reflects improvements in health care and primary prevention. The significant increases in the LMI Southeast Asia and Western Pacific regions reflect an aging population, increasing exposure to lifestyle risk factors and more effective treatment of other health burdens. During 2012, the LMI Southeast Asia region had a much larger proportion of men than women affected by IHD [28]. The significant gender difference in Southeast Asia is likely associated with a significant disparity in tobacco smoking [1].

Worldwide IHD estimated mortality rates demonstrate a different distribution; it is currently the commonest specific cause of death, at 13.2% [4]. Nowbar et al. performed one of the most recent geographical analyses of the global mortality burden of IHD. Fig. 4.10 demonstrates a cartogram of IHD mortality in 2010. A cartogram is a world atlas where countries' areas are algorithmically altered to represent the variable being measured (ie, IHD mortality). The larger the size of the country on the cartogram, the higher the actual mortality rate. The age-standardized mortality rates are represented by the shades of color in Fig. 4.10.

Nowbar et al. utilized data from the WHO to perform the IHD mortality rate analysis. Actual mortality rates were only available for 71 member states through data registration. The three countries with the highest IHD mortality were Russia, the United States, and Ukraine. Nowbar et al. utilized two international public health centers' (WHO and the Institute for Health Metrics and Evaluation) best statistical and epidemiological modeling to establish realistic estimates of IHD mortality. The 2012 IHD mortality data place India, China, Russia, United States, and Ukraine, in descending order, as the countries with highest global mortality [74].

Atrial Fibrillation

AF is the commonest cardiac arrhythmia worldwide [75]. AF has a prevalence in the general population of North America and Europe of 1–2% [76]. As the

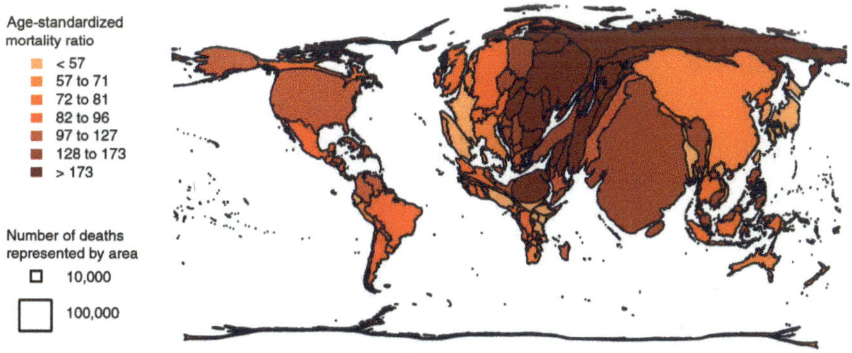

FIGURE 4.10 Cartogram showing the worldwide distribution of ischemic heart disease mortality using 2010 estimates [74].

population ages the AF prevalence increases from 0.5% aged 50–59 years going up to almost 9% at age 80–89 years [75]. The proportion of the population in North America and Europe with AF is estimated to double in the next 50 years [77]. Additionally, it is estimated that 3–6% of acute hospital admissions may be associated with AF [77].

The global burden of AF alone is very difficult to estimate as the bulk of the statistics are based on epidemiological studies from North America and Europe [77]. The population recruited for these studies are mainly white Caucasians. There is variation amongst different ethnic groups. In the United Kingdom, the West Birmingham-AF project demonstrated a prevalence rate of 2.4% in white Caucasians [78]. In contrast the prevalence rate was lower amongst in Indo-Asians, at 0.6% [78]. Furthermore, other studies suggest the proportion of AF in the black populations is lower [75]. Recognizing the lack of epidemiological studies on AF outside North American and Europe, Lip et al. performed a systematic review to identify and collate the current evidence [75]. The majority studies identified are from Japan, China, and Taiwan with the others being from Australia, Singapore, Thailand, Brazil, India, and South Korea [75]. The estimated prevalence of AF ranged from 0.1% in India to 4% in Australia [75]. The prevalence in Japan was fairly across all the studies included in the analysis of 0.6–1.6% [75]. The Chinese and Thai community cohorts were more variable, ranging from 0.8% to 2.8% and 0.4–2.2%, respectively [75].

Lip et al. demonstrated that in the majority of studies increasing age was directly associated with AF prevalence [75]. The only exception was within the Taiwan community-based cohort where the largest proportion with AF were in the age range 45–64 years [75]. Valvular heart disease is estimated to be the cause of AF into up to 22% of cases [75]. Hypertension was a common comorbidity in most of the studies included. Coexisting cardiac disease and diabetes mellitus where commonly reported comorbidities in China, Japan, South Korea, and Taiwan community studies [78]. The systematic review demonstrated the difference in AF between different world regions and the influence this has on disease burden globally [75].

Heart Failure

Heart failure (HF) is a heterogeneous condition that affects an estimated 26 million people worldwide [79]. It affects 800,000 people in the United Kingdom and 5.7 million people in the United States [79,80]. Furthermore the incidence of HF in the United States is 670,000 new cases per year [81]. HF is associated with a high mortality with an estimated 30–40% mortality rate within the first year of diagnosis [82]. However, there is an improving trend in mortality demonstrated by the six-month mortality rate decreasing from 26% in 1995 to 14% in 2005 [83]. The hospitalization rate for HF patients is high. In Europe and the United States, HF is a leading cause of hospitalization [79]. This accounts for over one million hospital admissions, which represent 1–2% of all hospitalizations [84]. In the United Kingdom, 2% of all National Health Service inpatient bed days and 5% of all emergency medical admissions are accounted for by HF. It is estimated that hospital admissions due to heart failure will rise by 50% over the next 25 years [32,85]. The age-adjusted hospitalization rate has decreased by 1–1.5% per annum since 1992/1993 [86]. The improvement rate in hospitalizations is due to more effective treatments, however, the burden of the aging population and improved survival from heart failure means that prevalence of the condition will continue to increase.

A global perspective on HF is challenging to establish. Many CVD conditions can lead to HF, including IHD, hypertensive heart disease, cardiomyopathy, congenital heart disease (CHD), rheumatic fever, and other conditions either alone or alongside other factors. The incidence and prevalence of HF worldwide will be different based on the exposure of risk factors. For instance, IHD is the predominant cause in high-income nations and nonischemic cardiomyopathy and rheumatic heart disease are more common in developing countries [87]. The burden of HF has changed in many regions with the advent of medical and device therapy. The added burden of many more patients surviving the causative condition means the prevalence continues to increase, and patients continue to survive longer depending on the geographical region. The great challenge with estimating global HF burden is the lack of uniformity in making the diagnosis across the world. ICD-10 does not clearly define HF as a separate diagnosis and is the reason HF is not examined in the GBD study [13,15]. HF definition is not applied uniformly even in high-income regions [88]. The result is that comparable statistics are not easily reproducible [89]. Some previous reviews have only focused on particular regions [89].

Ambrosy et al. reviewed hospital-based registries to gain real-world data on HF hospital admissions to examine clinical characteristics, regional variations, and limitations [79]. Clinical trials were not used due to variation in the eligibility criteria on severity, etiology, management, and outcomes [79]. Complete vital registration was limited due to the ICD classification of HF [13]. The majority of registries reviewed were North American or European, therefore complete global representation was not possible. Ambrosy et al. demonstrated that the overall mean age of patients admitted to hospital was 70–75 years [79]. Mean age difference amongst different regions maybe explained by the difference in prevalence of HF risk factors [79]. Interestingly the Acute Decompensation Heart Failure National Registry International–Asia Pacific (ADHERE-AP), which includes registry data for eight countries, demonstrated significant variation in age of presentation. Furthermore, the age of presentation per country was inversely associated with the human index, a composite score including life expectancy, adult literacy, educational level, and standard of living [90]. Interestingly the registries demonstrate there is virtually no data on ethnicity of HF hospital admissions outside the United States, and this data has its limits in terms of the quality of its recording [79]. In the US registries African Americans and Hispanics account for 20% and 7%, respectively, of HF hospital admissions

[79]. The ethnic groups present to hospital at a younger age and have lower ejection fractions [79]. Despite these factors, there appears to be no difference in their outcomes.

Khatibzadeh et al. performed a large systematic review and pooled analysis on studies on risk factors for HF to establish worldwide prevalence of major risk factors, which included 37 clinic-based studies representing 40 countries [89]. Global representation of HF studies was achieved, but the majority were based in Western Europe and North America [89]. Importantly, much of North Africa, the Middle East, India, and Australasia are not represented at all [89]. The Central and Eastern Europe population studies had the oldest population and the Sub-Saharan Africa studies had the youngest HF patients [89]. Furthermore, IHD and hypertension were reported separately in all studies, however, valvular heart disease, cardiopulmonary disease, and cardiomyopathy were not reported separately in all studies [89].

IHD as a risk factor was the commonest risk in Western, Eastern, and Central Europe. In Sub-Saharan Africa it accounts for <10% of risk factors. Hypertension is a commonly reported risk factor in all regions and has a crude global prevalence of 17% [89]. Applying age and sex adjustment for the prevalence, rates were significantly higher for hypertension as a risk factor in Eastern and Central Europe and Sub-Saharan Africa [89]. Fig. 4.11 demonstrates the age- and sex-adjusted relative contributions of different underlying risk factors of heart failure in six world regions [89].

Rheumatic heart disease was not reported in all selected publications. In those that reported rheumatic heart disease, the highest median crude prevalence is 34% in East Asia and 14% in Sub-Saharan Africa [89]. Cardiomyopathy was a major risk factor in Sub-Saharan Africa, Latin America, the Caribbean, and Asia Pacific region (Fig. 4.11) [89]. Importantly, cardiomyopathy was often attributed to Chagas disease in Latin American studies [91]. Khatibzadeh et al. make the observation that many HF cases were associated with multiple factors [91]. Under the ICD-10 coding rules a mortality due to HF must be aligned to one cause, although this analysis would assist in epidemiology modeling, attributing an HF death to one cause does not account for the many multifactorial cases [89].

Congenital Heart Disease

Congenital heart disease (CHD) or congenital heart anomalies is a significant cause of morbidity and mortality. The definition of CHD is a clinically significant structural heart diseased present at birth [92]. Strictly, the definition does not include abnormal vena caval drainage that has no clinical relevance. Moreover there are some genetic problems that occur after infancy that are not normally included; an example would be valve lesions due to Marfan syndrome or obstruction due to hypertrophic cardiomyopathy. The genetic cause of heart disease contributes little to the overincidence figures [92].

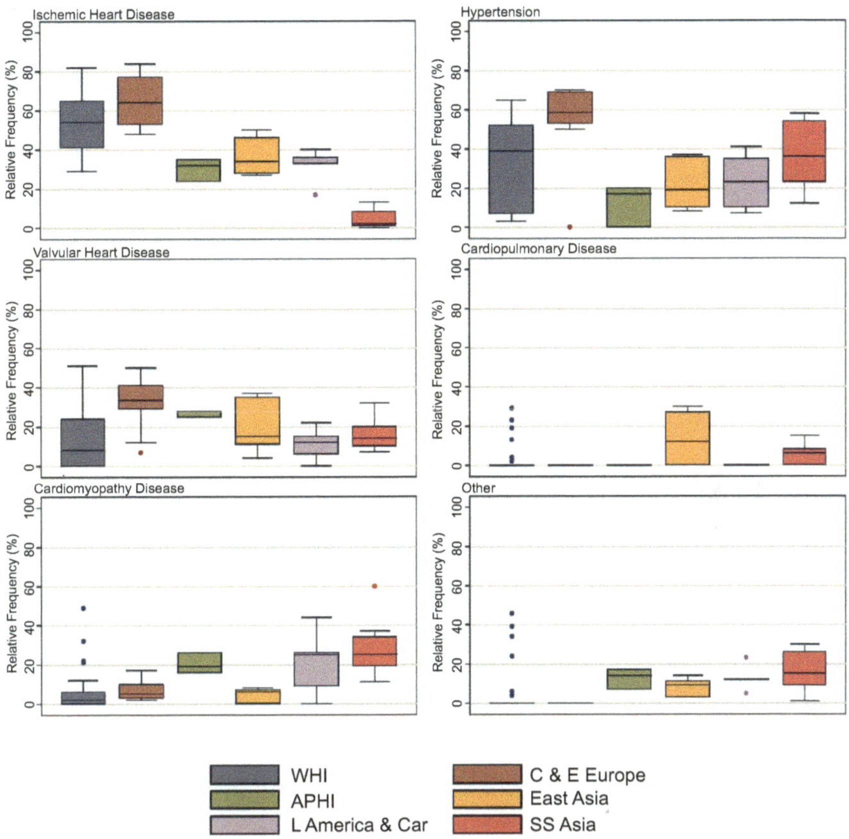

FIGURE 4.11 Age and sex adjusted relative contributions of different underlying risk factors of heart failure in six world regions [89].

Overall, CHD is a major cause of morbidity and mortality worldwide. In 2012 an estimated 230,000 deaths were attributable to CHD worldwide [4]. In 2000 the number of estimated deaths was higher at 234,000, but both years the proportion remained the same at 0.4% [4]. CHD in 2012 was the cause of an estimated 19.8 million lost DALYs, which was a reduction in 2000 from 20.3 million [8]. Proportionally this remained the same in both years at 0.7% [8]. Yet the incidence of CHD has increased with the ability to diagnose milder forms of the disease. The current incidence is an estimated 10–12 per 1000 live births [92]. These figures do not appear congruous with the decline in absolute numbers of mortality and loss of DALYs globally. Between 1990 and 2010 there was a reduction in loss of DALYs [2]. Globally these numbers reflect a greater rate of diagnosis, reflecting better screening and diagnosis. Many of the defects identified are mild and do not cause any long-term problems [92]. Many defects identified like small muscular ventricular septal defects will resolve in childhood [92].

TABLE 4.6 Live Births Per Total Population 2012 [92]

	Population	Birth	Births/ Population %
Global	7,075,456,000	138,584,400	2.0
Africa	894,247,000	33,544,900	3.8
Region of the Americas	956,779,000	15,569,400	1.6
Southeast Asia	1,833,361,000	36,504,300	2.0
Europe	904,484,000	11,215,800	1.2
Eastern Mediterranean	610,653,000	15,652,800	2.6
Western Pacific	1,844,750,000	25,706,000	1.4

The incidence of CHD remains fairly consistent globally. There are minor regional variations, for instance, Asian countries have a lower estimated incidence than elsewhere of coarctation of the aorta and aortic stenosis [92]. The biggest influence on global incidence of CHD is actually how many children are born per million population in a particular country or region. The global live birth rate for 2012 was 2.0%. Between regions this varied quite widely with Europe having the lowest live birth rate and Africa having the highest percentage [4]. Table 4.6 demonstrated the live birth rate percentages compared to total population for the WHO member states. The high live birth rate will represent a disproportionate number of children with CHD [92].

The countries with the higher birth rate tend to be those with lower per capita income, meaning that more resources are required in countries where resources are scarcer. Local costs in these countries are lower and this offsets this disparity slightly. The resources to treat CHD are inadequate and poorly distributed [93]. It is estimated that 75% of the world's population have no access to cardiac surgery, furthermore, the distribution of cardiac surgeons remains unbalanced [93]. For example, in Europe the ratio of surgeon to population is 1:3.5 million compared to Africa where the ratio is 1:38 million [93]. Resources are critical to redressing this balance, but the situation remains logistically and economically challenging. Many strategies are trying to redress this balance; firstly, certain patients will travel to where the treatment is available, but this is only an option for a few. Another option is taking the health care and pediatric cardiac surgical teams out to areas with high CHD burden; this, however, remains a challenge. Addressing the larger socioeconomic issues will remain the route to improving care in low-income countries.

CONCLUSIONS

CVD remains the most important NCD health burden worldwide, and this is unlikely to change over the coming decades. However, the pattern of CVD is likely to evolve, particularly in developing countries. Africa is likely to follow the trends seen in the Indian Sub-continent with increasing urbanization influencing physical activity and diet as well as improvement in management of communicable diseases; this will lead to a greater number and relative proportion of coronary events.

The impending worldwide epidemics of diabetes mellitus and AF will undoubtedly have significant roles to play, impacting heavily on the global burden of coronary disease and heart failure, cementing CVD as the most significant global health burden for decades to come.

ACKNOWLEDGMENTS

The authors would like the library team at University Hospital Coventry and Warwickshire NHS Trust for their assistance with performing the literature review. Special thanks in particular to Petra Meeson for her advice, assistance, and hard work.

REFERENCES

[1] Mendis S, Puska P, Norrving B. Global atlas on cardiovascular disease prevention and control. World Health Organization; 2011.

[2] Moran AE, Roth GA, Narula J, Mensah GA. 1990–2010 global cardiovascular disease atlas. Glob Heart 2014;9(1):3–16.

[3] Laslett LJ, Alagona Jr P, Clark 3rd BA, Drozda Jr JP, Saldivar F, Wilson SR, et al. The worldwide environment of cardiovascular disease: prevalence, diagnosis, therapy, and policy issues: a report from the American College of Cardiology. J Am Coll Cardiol 2012;60(Suppl. 25):S1–49.

[4] World Health Organization. Global health estimates 2014 summary tables: Deaths by cause, age and sex, by WHO region, 2000–2012(Internet). 2014. Available from: http://www.who.int/healthinfo/global_burden_disease/en/ (accessed 23.03.15).

[5] Fuster V. Global burden of cardiovascular disease: time to implement feasible strategies and to monitor results. J Am Coll Cardiol 2014;64(5):520–2.

[6] Moran A, Forouzanfar M, Sampson U, Chugh S, Feigin V, Mensah G. The epidemiology of cardiovascular diseases in Sub-Saharan Africa: the global burden of diseases, injuries and risk factors. Study Progress Cardiovasc Dis 2010;56(3):234–9.

[7] World Health Organization. Global status report on noncommunicable diseases 2014. 2014.

[8] World Health Organization. Global health estimates 2014 summary tables: DALY by cause, age and sex, by WHO region, 2000–2012 (Internet). 2014. Available from: http://www.who.int/healthinfo/global_burden_disease/en/ (accessed 18.04.15).

[9] Vos T, Flaxman AD, Naghavi M, Lozano R, Michaud C, Ezzati M, et al. Years lived with disability (YLDs) for 1160 sequelae of 289 diseases and injuries 1990–2010: a systematic analysis for the Global Burden of Disease Study 2010. Lancet 2012;380(9859):2163–96.

[10] GBD 2013 Mortality and Causes of Death Collaborators, Global, regional, and national age-sex specific all-cause and cause-specific mortality for 240 causes of death, 1990–2013: a systematic analysis for the Global Burden of Disease Study 2013. Lancet 2015;385(9963):117–71.

[11] World Health Organization. Global health estimates 2014 summary tables: Deaths by cause, age and sex, by World Bank region, 2000–2012 (Internet). 2014. Available from: http://www.who.int/healthinfo/global_burden_disease/en/ (accessed 03.04.15).

[12] World Health Organization. Global health estimates 2014 summary tables: Estimated deaths ('000) by cause, sex and WHO member state [Internet]. 2014. Available from: http://www.who.int/healthinfo/global_burden_disease/en/ (accessed 01.05.15).

[13] World Health Organization. International statistical classification of diseases and related health problems. World Health Organization; 2004.
[14] Mathers CSG, Ma Fat D, Ho J, Mahanani WR. WHO methods and data sources for country – level causes of death 2000–2012. Department of Health Statistics and Information Systems, WHO G; 2014 May 2014. Report No.: Contract No.: Global Health Estimates Technical Paper WHO/HIS/HSI/GHE/2014.7.
[15] Murray CJ, Ezzati M, Flaxman AD, Lim S, Lozano R, Michaud C, et al. GBD 2010: design, definitions, and metrics. Lancet 2012;380(9859):2063–6.
[16] Pagidipati NJ, Gaziano TA. Estimating deaths from cardiovascular disease: a review of global methodologies of mortality measurement. Circulation 2013;127(6):749–56.
[17] Keys A, Taylor HL, Blackburn H, Brozek J, Anderson JT, Simonson E. Coronary heart disease among Minnesota business and professional men followed fifteen years. Circulation 1963;28:381–95.
[18] Roulson J, Benbow EW, Hasleton PS. Discrepancies between clinical and autopsy diagnosis and the value of post mortem histology; a meta-analysis and review. Histopathology 2005;47(6):551–9.
[19] Modelmog D, Rahlenbeck S, Trichopoulos D. Accuracy of death certificates: a population-based, complete-coverage, one-year autopsy study in East Germany. Cancer Causes Control 1992;3(6):541–6.
[20] Ravakhah K. Death certificates are not reliable: revivification of the autopsy. South Med J 2006;99(7):728–33.
[21] Setel PW, Macfarlane SB, Szreter S, Mikkelsen L, Jha P, Stout S, et al. A scandal of invisibility: making everyone count by counting everyone. Lancet 2007;370(9598):1569–77.
[22] Yang G, Hu J, Rao KQ, Ma J, Rao C, Lopez AD. Mortality registration and surveillance in China: history, current situation and challenges. Popul Health Metr 2005;3(1):3.
[23] Jha P, Gajalakshmi V, Gupta PC, Kumar R, Mony P, Dhingra N, et al. Prospective study of one million deaths in India: rationale, design, and validation results. PLoS Med 2006;3(2):e18.
[24] Moran AE, Forouzanfar MH, Roth GA, Mensah GA, Ezzati M, Flaxman A, et al. The global burden of Ischemic heart disease in 1990 and 2010: the global burden of disease 2010 study. Circulation 2014;129(14):1493–501.
[25] World Health Organization. Global health estimated 2014 summary tables [Internet]. 2014. Available from: http://www.who.int/healthinfo/global_burden_disease/en/ (accessed 05.04.15).
[26] Nascimento BR, Brant LCC, Moraes DN, Ribeiro ALP. Global health and cardiovascular disease. Heart 2014;100(22):1743–9.
[27] Baena CP, Chowdhury R, Schio NA, Sabbag Jr AE, Guarita-Souza LC, Olandoski M, et al. Ischemic heart disease deaths in Brazil: current trends, regional disparities and future projections. Heart 2013;99(18):1359–64.
[28] World Health Organization. Global health estimates 2014 summary tables: DALY by cause, age and sex, by World Bank income category and WHO region 2000–2012 (Internet). 2014. Available from: http://www.who.int/healthinfo/global_burden_disease/en/ (accessed 11.05.15).
[29] Alwan A. Global status report on noncommunicable diseases 2010. World Health Organization; 2011.
[30] World Health Organization. Global health risks: Mortality and burden of disease attributable to selected major risks. World Health Organization; 2009.
[31] Danaei G, Finucane MM, Lin JK, Singh GM, Paciorek CJ, Cowan MJ, et al. National, regional, and global trends in systolic blood pressure since 1980: systematic analysis of health examination surveys and epidemiological studies with 786 country-years and 5.4 million participants. The Lancet 2011;377(9765):568–77.

[32] Townsend NWK, Bhatnagar P, Smolina K, Nichols MLM, Luengo-Fernandez R, Rayner M. Coronary heart disease statistics: Heart failure supplement. London: British Heart Foundation; 2002.

[33] Wild S, Roglic G, Green A, Sicree R, King H. Global prevalence of diabetes: estimates for the year 2000 and projections for 2030. Diabet Care 2004;27(5):1047–53.

[34] Levitan EB, Song Y, Ford ES, Liu S. Is nondiabetic hyperglycemia a risk factor for cardiovascular disease?: a meta-analysis of prospective studies. Arch Intern Med 2004;164(19):2147–55.

[35] World Health Organization. UNAIDS. Prevention of cardiovascular disease. World Health Organization; 2007.

[36] Fantoni C, Regoli F, Ghanem A, Raffa S, Klersy C, Sorgente A, et al. Long-term outcome in diabetic heart failure patients treated with cardiac resynchronization therapy. Eur J Heart Fail 2008;10(3):298–307.

[37] Smooke S, Horwich TB, Fonarow GC. Insulin-treated diabetes is associated with a marked increase in mortality in patients with advanced heart failure. Am Heart J 2005;149(1):168–74.

[38] Shindler DM, Kostis JB, Yusuf S, Quinones MA, Pitt B, Stewart D, et al. Diabetes mellitus, a predictor of morbidity and mortality in the Studies of Left ventricular Dysfunction (SOLVD) trials and registry. Am J Cardiol 1996;77(11):1017–20.

[39] De Groote P, Lamblin N, Mouquet F, Plichon D, McFadden E, Van Belle E, et al. Impact of diabetes mellitus on long-term survival in patients with congestive heart failure. Eur Heart J 2004;25(8):656–62.

[40] Tenenbaum A, Motro M, Fisman EZ, Leor J, Freimark D, Boyko V, et al. Functional class in patients with heart failure is associated with the development of diabetes. Am J Med 2003;114(4):271–5.

[41] Shafey O, Eriksen M, Ross H, Mackay J. The tobacco atlas. Atlanta: American Cancer Society; 2009.

[42] World Health Organization. WHO report on the global tobacco epidemic, 2011: Warning about the dangers of tobacco. Geneva: World Health Organization; 2011.

[43] Yusuf S, Hawken S, Ounpuu S, Dans T, Avezum A, Lanas F, et al. Effect of potentially modifiable risk factors associated with myocardial infarction in 52 countries (the INTERHEART study): case-control study. Lancet 2004;364(9438):937–52.

[44] Sattelmair J, Pertman J, Ding EL, Kohl HW, Haskell W, Lee I-M. Dose response between physical activity and risk of coronary heart disease a meta-analysis. Circulation 2011;124(7):789–95.

[45] Lopez AD, Murray C. The global burden of disease. Nat Med 1998;4(11):1241–3.

[46] Shestov DB, Deev AD, Klimov AN, Davis CE, Tyroler HA. Increased risk of coronary heart disease death in men with low total and low-density lipoprotein cholesterol in the Russian Lipid Research Clinics Prevalence Follow-up Study. Circulation 1993;88(3):846–53.

[47] Broda G, Davis CE, Pajak A, Williams OD, Rywik SL, Baczyńska E, et al. Poland and United States collaborative study on cardiovascular epidemiology: a comparison of HDL cholesterol and its subfractions in populations covered by the United States atherosclerosis risk in communities study and the Pol-MONICA project. Arterioscler Thromb Vasc Biol 1996;16(2):339–49.

[48] Britton A, McKee M. The relation between alcohol and cardiovascular disease in Eastern Europe: explaining the paradox. J Epidemiol Community Health 2000;54(5):328–32.

[49] Chenet L, McKee M, Leon D, Shkolnikov V, Vassin S. Alcohol and cardiovascular mortality in Moscow; new evidence of a causal association. J Epidemiol Community Health 1998;52(12):772–4.

[50] Evans C, Chalmers J, Capewell S, Redpath A, Finlayson A, Boyd J, et al. "I don't like Mondays"—day of the week of coronary heart disease deaths in Scotland: study of routinely collected data. BMJ 2000;320(7229):218–9.

[51] Yusuf S, Reddy S, Ounpuu S, Anand S. Global burden of cardiovascular diseases: part I: general considerations, the epidemiologic transition, risk factors, and impact of urbanization. Circulation 2001;104(22):2746–53.
[52] World Health Organization. Life expectancy: Data by WHO region [Internet]. 2014. Available from: http://apps.who.int/gho/data/view.main.690?lang=en (accessed 12.05.15).
[53] Baker J, Mitchell R, Lawson K, Pell J. Ethnic differences in the cost-effectiveness of targeted and mass screening for high cardiovascular risk in the UK: cross-sectional study. Heart 2013;99(23):1766–71.
[54] Yusuf S, Reddy S, Ounpuu S, Anand S. Global burden of cardiovascular diseases: Part II: variations in cardiovascular disease by specific ethnic groups and geographic regions and prevention strategies. Circulation 2001;104(23):2855–64.
[55] Artaud-Wild SM, Connor SL, Sexton G, Connor WE. Differences in coronary mortality can be explained by differences in cholesterol and saturated fat intakes in 40 countries but not in France and Finland. A paradox. Circulation 1993;88(6):2771–9.
[56] McGovern PG, Pankow JS, Shahar E, Doliszny KM, Folsom AR, Blackburn H, et al. Recent trends in acute coronary heart disease–mortality, morbidity, medical care, and risk factors. The Minnesota Heart Survey Investigators. N Engl J Med 1996;334(14):884–90.
[57] Vartiainen E, Puska P, Pekkanen J, Tuomilehto J, Jousilahti P. Changes in risk factors explain changes in mortality from ischemic heart disease in Finland. BMJ 1994;309(6946):23–7.
[58] Thorvaldsen P, Asplund K, Kuulasmaa K, Rajakangas AM, Schroll M. Stroke incidence, case fatality, and mortality in the WHO MONICA project. World Health Organization monitoring trends and determinants in cardiovascular disease. Stroke 1995;26(3):361–7.
[59] Chen Z, Peto R, Collins R, MacMahon S, Lu J, Li W. Serum cholesterol concentration and coronary heart disease in population with low cholesterol concentrations. BMJ 1991;303(6797):276–82.
[60] Shah BR, Victor JC, Chiu M, Tu JV, Anand SS, Austin PC, et al. Cardiovascular complications and mortality after diabetes diagnosis for South Asian and Chinese patients: a population-based cohort study. Diabet Care 2013;36(9):2670–6.
[61] Gupta M, Singh N, Verma S. South Asians and cardiovascular risk: what clinicians should know. Circulation 2006;113(25):e924–9.
[62] Blackledge HM, Newton J, Squire IB. Prognosis for South Asian and white patients newly admitted to hospital with heart failure in the United Kingdom: historical cohort study. BMJ 2003;327(7414):526–31.
[63] Anand SS, Yusuf S, Vuksan V, Devanesen S, Teo KK, Montague PA, et al. Differences in risk factors, atherosclerosis and cardiovascular disease between ethnic groups in Canada: the study of health assessment and risk in ethnic groups (SHARE). Indian Heart J 2000;52 (Suppl. 7):S35–43.
[64] Sheth T, Nair C, Nargundkar M, Anand S, Yusuf S. Cardiovascular and cancer mortality among Canadians of European, south Asian and Chinese origin from 1979 to 1993: an analysis of 1.2 million deaths. CMAJ 1999;161(2):132–8.
[65] Pais P, Pogue J, Gerstein H, Zachariah E, Savitha D, Jayprakash S, et al. Risk factors for acute myocardial infarction in Indians: a case-control study. Lancet 1996;348(9024):358–63.
[66] Ramachandran A, Snehalatha C, Dharmaraj D, Viswanathan M. Prevalence of glucose intolerance in Asian Indians. Urban-rural difference and significance of upper body adiposity. Diabetes Care 1992;15(10):1348–55.
[67] Gupta R, Gupta VP. Meta-analysis of coronary heart disease prevalence in India. Indian Heart J 1996;48(3):241–5.
[68] Passel JS, Cohn D, Lopez MH. Hispanics account for more than half of nation's growth in past decade. Pew Hispanic Center; 2011. http://pewhispanicorg/files/reports/140pdf.

[69] Sorlie PD, Aviles-Santa LM, Wassertheil-Smoller S, Kaplan RC, Daviglus ML, Giachello AL, et al. Design and implementation of the hispanic community health study/study of Latinos. Ann Epidemiol 2010;20(8):629–41.

[70] Daviglus ML, Talavera GA, Aviles-Santa ML, Allison M, Cai J, Criqui MH, et al. Prevalence of major cardiovascular risk factors and cardiovascular diseases among Hispanic/Latino individuals of diverse backgrounds in the United States. JAMA 2012;308(17):1775–84.

[71] Berrios X, Koponen T, Huiguang T, Khaltaev N, Puska P, Nissinen A. Distribution and prevalence of major risk factors of noncommunicable diseases in selected countries: the WHO Inter-Health Programme. Bull World Health Organ 1997;75(2):99–108.

[72] Marmot MG, Syme SL, Kagan A, Kato H, Cohen JB, Belsky J. Epidemiologic studies of coronary heart disease and stroke in Japanese men living in Japan, Hawaii and California: prevalence of coronary and hypertensive heart disease and associated risk factors. Am J Epidemiol 1975;102(6):514–25.

[73] Takeya Y, Popper JS, Shimizu Y, Kato H, Rhoads GG, Kagan A. Epidemiologic studies of coronary heart disease and stroke in Japanese men living in Japan, Hawaii and California: incidence of stroke in Japan and Hawaii. Stroke 1984;15(1):15–23.

[74] Nowbar AN, Howard JP, Finegold JA, Asaria P, Francis DP. Global geographic analysis of mortality from ischemic heart disease by country, age and income: statistics from World Health Organisation and United Nations. Int J Cardiol 2014;174(2):293–8.

[75] Lip GYH, Brechin CM, Lane DA. The global burden of atrial fibrillation and stroke: a systematic review of the epidemiology of atrial fibrillation in regions outside North America and Europe. Chest 2012;142(6):1489–98.

[76] Stewart S, Hart CL, Hole DJ, McMurray JJ. Population prevalence, incidence, and predictors of atrial fibrillation in the Renfrew/Paisley study. Heart 2001;86(5):516–21.

[77] Miyasaka Y, Barnes ME, Gersh BJ, Cha SS, Bailey KR, Abhayaratna WP, et al. Secular trends in incidence of atrial fibrillation in Olmsted County, Minnesota, 1980 to 2000, and implications on the projections for future prevalence. Circulation 2006;114(2):119–25.

[78] Lip GY, Bawden L, Hodson R, Rutland E, Snatchfold J, Beevers DG. Atrial fibrillation amongst the Indo-Asian general practice population. The West Birmingham atrial fibrillation project. Int J Cardiol 1998;65(2):187–92.

[79] Ambrosy AP, Fonarow GC, Butler J, Chioncel O, Greene SJ, Vaduganathan M, et al. The global health and economic burden of hospitalizations for heart failure: lessons learned from hospitalized heart failure registries. J Am Coll Cardiol 2014;63(12):1123–33.

[80] McMurray JJ, Adamopoulos S, Anker SD, Auricchio A, Bohm M, Dickstein K, et al. ESC guidelines for the diagnosis and treatment of acute and chronic heart failure 2012: the task force for the diagnosis and treatment of acute and chronic heart failure 2012 of the European society of Cardiology. Developed in collaboration with the heart failure association (HFA) of the ESC. Eur J Heart Fail 2012;14(8):803–69.

[81] Heidenreich PA, Albert NM, Allen LA, Bluemke DA, Butler J, Fonarow GC, et al. Forecasting the impact of heart failure in the United States: a policy statement from the American Heart Association. Circ Heart Fail 2013;6(3):606–19.

[82] Cowie MR, Wood DA, Coats AJ, Thompson SG, Suresh V, Poole-Wilson PA, et al. Survival of patients with a new diagnosis of heart failure: a population based study. Heart 2000;83(5):505–10.

[83] Mehta PA, Dubrey SW, McIntyre HF, Walker DM, Hardman SM, Sutton GC, et al. Improving survival in the 6 months after diagnosis of heart failure in the past decade: population-based data from the UK. Heart 2009;95(22):1851–6.

[84] Zannad F, Agrinier N, Alla F. Heart failure burden and therapy. Europace 2009;11 (Suppl. 5):v1–9.

[85] Stewart S, Horowitz JD. Home-based intervention in congestive heart failure: long-term implications on readmission and survival. Circulation 2002;105(24):2861–6.
[86] Mosterd A, Reitsma JB, Grobbee DE. Angiotensin converting enzyme inhibition and hospitalisation rates for heart failure in the Netherlands, 1980 to 1999: the end of an epidemic? Heart 2002;87(1):75–6.
[87] Mendez GF, Cowie MR. The epidemiological features of heart failure in developing countries: a review of the literature. Int J Cardiol 2001;80(2–3):213–9.
[88] Murdoch DR, Love MP, Robb SD, McDonagh TA, Davie AP, Ford I, et al. Importance of heart failure as a cause of death. Changing contribution to overall mortality and coronary heart disease mortality in Scotland 1979–1992. Eur Heart J 1998;19(12):1829–35.
[89] Khatibzadeh S, Farzadfar F, Oliver J, Ezzati M, Moran A. Worldwide risk factors for heart failure: a systematic review and pooled analysis. Int J Cardiol 2013;168(2):1186–94.
[90] Atherton JJ, Hayward CS, Wan Ahmad WA, Kwok B, Jorge J, Hernandez AF, et al. Patient characteristics from a regional multicenter database of acute decompensated heart failure in Asia Pacific (ADHERE International-Asia Pacific). J Card Fail 2012;18(1):82–8.
[91] Castro P, Vukasovic JL, Garces E, Sepulveda L, Ferrada M, Alvarado S, et al. Cardiac failure in Chilean hospitals: results of the national registry of heart failure, ICARO. Rev Med Chil 2004;132(6):655–62.
[92] Hoffman J. The global burden of congenital heart disease. Cardiovasc J Afr 2013;24(4):141–5.
[93] Bernier PL, Stefanescu A, Samoukovic G, Tchervenkov CI. The challenge of congenital heart disease worldwide: epidemiologic and demographic facts. Semin Thorac Cardiovasc Surg Pediatr Card Surg Annu 2010;13(1):26–34.

Chapter 5

Lifestyle Choices, Risk Factors, and Cardiovascular Disease

S. Chiesa[1], E. Farmaki[2], E. Zacharia[2], D. Tousoulis[2], M. Charakida[1]
[1]University College London, London, United Kingdom; [2]Athens University Medical School, Athens, Greece

List of Abbreviations

BMI Body mass index
CHD Coronary heart disease
CRP C-reactive protein
CV Cardiovascular
CVD Cardiovascular disease
DALY Disability-adjusted life year
eNOS Endothelial nitric oxide synthase
EPC Endothelial progenitor cell
FMD Flow-mediated dilation
GDP Gross Domestic Product
GI Glycemic index
GL Glycemic load
HDL-C High-density-lipoprotein cholesterol
IL-6 Interleukin-6
LDL-C Low-density-lipoprotein cholesterol
MCP-1 Monocyte chemoattractant protein 1
MUFA Monounsaturated fatty acid
NO Nitric oxide
ONTARGET Ongoing Telmisartan Alone and in Combination with Ramipril Global Endpoint Trial
PAI-1 Plasminogen activator inhibitor 1
PAF Platelet activating factor
PUFA Polyunsaturated fatty acid
RH-PAT Reactive hyperemia peripheral arterial tonometry
SFA Saturated fatty acid
TMAO Trimethylamine-N-oxide
TNF-α Tumor necrosis factor alpha
TRANSCEND Telmisartan Randomized Assessment Study in ACE Intolerant Subjects with Cardiovascular Disease
US United States

Cardiovascular Diseases.

INTRODUCTION

Cardiovascular disease (CVD) remains one of the leading causes of mortality in both the developed and developing world, accounting for over 17.5 million deaths in 2012 alone [1]. In addition to premature death, CVD results in a substantial burden of disability for individuals living with the disease, with estimated costs totaling 3% of Gross Domestic Product (GDP) in the United States and leading to an annual financial burden in excess of €192 billion in the European Union. Although genetic factors such as aging, gender, and family history contribute to the development of CVD, it has been estimated that over 90% of the population attributable risk for myocardial infarction alone can be attributed to a number of modifiable risk factors that act to significantly increase atherosclerotic risk [2]. Lifestyle behaviors responsible for these risk factors are commonly found in high-income, highly urbanized countries and are characterized by unhealthy dietary patterns, low levels of physical activity, excessive tobacco and alcohol consumption, and societal and economic factors such as social deprivation and lack of education. These behaviors act both in isolation and in synergy to predispose individuals to a multitude of well-established cardiovascular (CV) risk factors, and these resulting risk factors in turn interact with each other to further elevate CVD risk (Fig. 5.1). In this

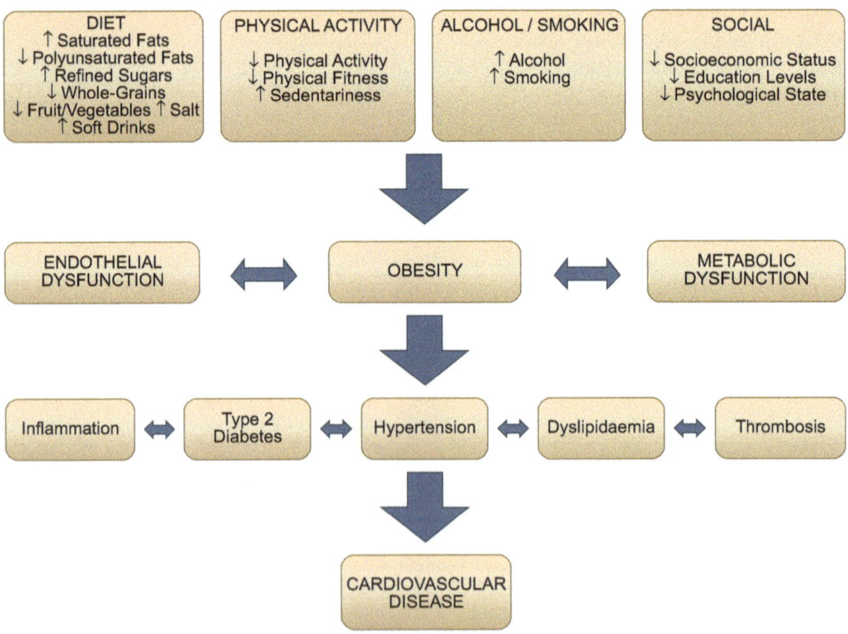

FIGURE 5.1 Relationships between lifestyle choices, risk factors, and CVD. The adoption of unhealthy lifestyle choices results in the development of multiple risk factors for CVD. These factors act both in isolation and in synergy to significantly increase atherosclerotic progression and subsequent risk of CV morbidity and mortality.

chapter, we will document how different lifestyle choices can affect CV health, and provide evidence of how these can be modified by lifestyle interventions.

DIETARY PATTERNS AND CV RISK

The recent emergence of intensive agriculture practices and the expansion of ubiquitous food and drink brands as part of the global fast-food phenomenon have resulted in changes in global dietary patterns that are without doubt the most dramatic in human history. The impact of poor dietary habits on worldwide morbidity and mortality is staggering, with up to 16.3 million deaths a year attributed to causes directly related to six distinct dietary risk factors: low fruit intake, high sodium intake, low nut or seed intake, low vegetable intake, low whole-grain intake, and low seafood omega-3 fatty acid intake [3].

The majority of studies linking diet and CVD risk are based on cross-sectional, prospective, and case–control studies. Although convincing evidence of associations is frequently provided from these forms of study, their observational nature prevents the establishment of causality, and the use of self-reporting questionnaires is prone to measurement error due to frequent underreporting of energy intake. Further problems arise due to the frequent clustering of lifestyle risk factors that may independently affect CVD outcomes, with people eating unhealthy diets more likely to be poorly educated, economically deprived, frequent drinkers and smokers, physically inactive, and suffering from other chronic diseases than those who eat well. Despite these difficulties, however, insights gained from a wealth of both observational and interventional studies over more than half a century of intensive research have attempted to shape dietary guidelines aimed at reducing CVD risk. While much early research focused on individual nutrients as risk factors for CVD, a growing appreciation of the importance of potential nutrient interactions and the synergistic effect of different food groups has expanded research to focus on both food groups as a whole and overall dietary patterns.

Individual Nutrients

Fats

The cornerstone of dietary guidelines over the past 40–50 years has centered on a reduction in dietary fat intake—specifically, fats composed of saturated and *trans*-fatty acids. While *trans*-fatty acids almost certainly increase the risk of CVD through factors such as their elevation of low-density-lipoprotein cholesterol (LDL-C), triglycerides, and total/HDL-C (low-density-lipoprotein cholesterol) ratio [4], the evidence surrounding saturated fatty acids (SFAs) and CVD remains unclear. Despite early ecological data showing a relationship between increased SFA consumption and CVD [5], more recent interventional studies and *meta*-analyses of cohort studies appear to show no relative CVD risk in individuals consuming increased levels of SFAs [6,7]. Care must be taken, however, in assessing the impact of SFA in the absence of information on overall dietary intake, as a lowered intake of this subgroup of fats implies that the intake of other nutrient groups must be increased in order to maintain daily calorific balance. In line with this, both observational and interventional studies in which energy intake from SFAs is replaced with polyunsaturated fatty acids (PUFAs) have consistently shown a 2–3% reduction in CVD risk for every 1% of SFA energy replaced by PUFA energy [8], supporting the argument for both a reduction in SFA intake and increase in PUFA intake as an effective means of reducing CVD risk. The argument for monounsaturated fatty acids (MUFAs) as a substitute for SFAs is less clear. While diets rich in MUFAs are clearly cardioprotective [9], pooled risk analyses of studies in which SFAs are substituted for MUFAs appear to indicate an increased risk for CVD events, despite positive effects on LDL-C concentration [10].

Carbohydrates

The effect of replacing SFA with carbohydrates is also complex, with CV outcomes dependent not on total carbohydrate intake *per se*, but rather on factors such as the type of carbohydrate consumed (eg, dietary fiber, whole-grain, or processed), the impact of the glycemic index/load (GI/GL) on postprandial insulin responses, and the structure of the carbohydrates when consumed (ie, solid vs. liquid). While carbohydrates consumed in the form of whole grains or dietary fibers are consistently associated with a reduced risk of multiple CVD risk factors [11,12], the high prevalence of refined and processed sugars commonly found in western-style diets may likely result in the replacement of SFAs with the arguably more detrimental high-GI/GL carbohydrate groups—further elevating CVD risk through excess weight gain, insulin resistance, postprandial inflammation, and endothelial dysfunction [13,14].

Sodium and Potassium

Excess sodium intake has been associated with increased blood pressure, endothelial dysfunction, vascular remodeling, albuminuria, and impaired renal function and CV outcome [15,16]. A meta-analysis of 13 prospective studies reported that 2000mg higher sodium intake was associated with a 14% increased risk for CVD [17]. Interestingly, dietary potassium supplementation appears to attenuate these adverse effects relevant to high dietary salt intake [15], and it is therefore plausible to speculate that modest dietary salt restriction accompanied by increasing potassium intake may serve as an effective strategy to prevent or control hypertension and decrease CV morbidity and mortality [18]. According to the US Department of Health and Human Services and US Department of Agriculture's 2010 Dietary Guidelines, a reduction of sodium intake to less than 2300mg/day per person with an even lower target of 1500mg/day for specific subpopulations is needed [19]. However, these recommendations have been heavily challenged [15]. In an observational analysis of the ONTARGET (Ongoing Telmisartan Alone and in

Combination with Ramipril Global Endpoint Trial) and TRANSCEND (Telmisartan Randomized Assessment Study in ACE Intolerant Subjects with Cardiovascular Disease) studies, a J-shaped relationship between urinary sodium excretion and risk of a composite outcome of CV events and mortality was noted [20]. These paradoxical findings may be explained by the difficulty in older subjects of deciphering whether CVD is the cause rather than the consequence of the level of salt intake. In addition, sick individuals are more likely to have adverse outcomes, and, as the severity of the illness progresses, caloric intake along with salt intake may dramatically decrease. Therefore, further studies with strict protocols are needed to provide information and guidance on optimum levels of sodium and potassium dietary intake at different ages.

Food Groups

Red Meat

Red meat is a major source of dietary protein and fat and has long been associated with an increased risk of both type 2 diabetes and CVD mortality [21]—with this risk significantly increased following the consumption of processed compared to unprocessed meat [22]. Mechanisms traditionally thought to be responsible for the increase in CVD risk are numerous, but include the high levels of saturated fat, cholesterol, heme iron, sodium, and nitrites found within this food group. Recently, the production of trimethylamine N-oxide (TMAO), a by-product of choline and carnitine (found in meat and dairy) and produced in part by the gut microbiome, has been proposed as a novel mechanism for CVD risk—indicating that differences in the gut microbiome resulting from distinct dietary patterns may impact CVD risk through previously unconsidered metabolic risk factors [23,24].

Fish

The regular consumption of fish has repeatedly been shown to reduce CV risk in a dose-dependent manner. Importantly, this relationship appears to be curvilinear, with changes from a no-fish diet to as little as a single portion a week providing the greatest reduction in risk. Pooled analyses estimate that eating one to two portions of fish a week may reduce the risk of CHD by up to 15% and the risk of stroke by 18% [25,26]. The major benefits of fish consumption are attributed to the presence of high concentrations of the omega-3 polyunsaturated acids, eicosapentaenoic acid and docosahexaenoic acid. These are found in higher quantities in oily fish such as salmon or herring, and as such, these types of fish appear more strongly related to reduced CV risk than lean fish [27].

Fruit and Vegetables

A diet low in fruit and vegetables is a significant risk factor for CV disease and has been estimated to account for 13.7% of the population attributable risk for myocardial infarction alone [2]. Health benefits associated with increased fruit and vegetable intake may result from a number of factors—including the replacement of less healthy food groups due to increased fruit intake or the composition of micronutrients contained within the fruits themselves. Fruit and vegetables are a major source of polyphenols, which have been shown to exert a cardioprotective effect through mechanisms such as improvements in endothelial function and reductions in oxidative stress [28,29]. In a recent meta-analysis of nine studies, endothelial function (measured by flow-mediated dilation (FMD)) improved within the first 120 min following intake of grape polyphenols, with more pronounced changes seen in subjects with high CV risk burden [30]. The impact of fruit polyphenols on CV risk factors is currently a focus of intense research, and studies into increased consumption of grape products alone have reported hypotensive, hypolipidemic, and atheroprotective properties.

Cocoa Products

Cocoa products, such as dark chocolate, are other sources of flavonoids that have gained widespread attention with regard to CV disease prevention. Antioxidant activity, reduction of platelet activation, enhancement of nitric oxide (NO) production, and blood pressure moderation are some of the favorable properties of these flavonoids [31,32]. A meta-analysis of 13 trials concluded that dark chocolate decreased systolic hypertension and diastolic prehypertension in a more effective manner than placebo [33]. These findings were also confirmed in a randomized trial in which both solid dark chocolate and liquid cocoa intake ameliorated endothelial function and decreased blood pressure in overweight adults compared to placebo [34]. In contrast, the nonflavonoid white chocolate was evidently inferior to the flavonoid-rich dark chocolate in terms of coronary circulation enhancement in healthy adults [35]. Although the existing data attributing a cardioprotective role to chocolate are encouraging, further understanding of the underlying mechanisms is needed before sugar- and saturated fat–rich chocolate products could be recommended as part of a daily diet [31].

Drinks

Soft Drinks

Data from multiple health surveys in the United States have shown that percentage caloric intake from sugar-sweetened beverages (soft drinks) has increased by almost 10% between 1965 and 2002, providing a considerable source of additional daily calories. As well as being nutrient deficient, the consumption of carbohydrates in liquid form appears to be less satiating than those ingested in solid form, leading to decreased dietary compensation during meals and an increased energy intake over time [36]. Regular consumption of soft drinks has been shown in prospective studies to increase the risk of adiposity and type 2 diabetes [37,38]. Conversely, interventional studies decreasing soft drink intake have reported decreases in multiple risk factors for CVD [39]. Recent research has also indicated that alongside long-term weight gain, regular soft drink consumption may be an independent risk factor for the development of type 2 diabetes even in the absence of increased adiposity [40].

Alcoholic Beverages

Moderate consumption of alcoholic beverages on a regular basis can reduce the risk of CHD by at least 40% [41–43]. Alcohol intake and risk of CHD are inversely correlated, but this advantageous effect exists only for low and moderate consumption, whereas high consumption of alcohol is associated with increased morbidity and mortality [44–46]. Still, consumption of up to two alcoholic beverages per day can independently improve endothelial function compared to abstention from alcohol [47]—an effect that has been reported in both healthy young subjects and patients with type 2 diabetes. The relatively low incidence of CV disease among the French population, known as "the French paradox," is likely to be the result of the prevalent consumption of red wine [48]. Next to alcohol, red wine contains a significant amount of polyphenols that can potently decrease CVD risk through multiple factors, including reductions in LDL-C and its susceptibility to oxidation, elevation of HDL-C, decreased platelet aggregation with parallel increases in fibrinolytic activity, and—last but not least—improvement of endothelial function by enhancement of NO production [48–52]. Red wine consumption of 400mL/day for 2weeks has been shown to significantly improve oxidative stress in both young and old populations [53], while the administration of a single dose of red wine with or without alcohol has shown a greater increase in FMD following the consumption of alcohol-free red wine—suggesting that the cardioprotective effect of red wine is more likely attributable to the antioxidant properties it possesses rather than the ethanol alone [54]. Accordingly, studies in hypertensive and normotensive adults have concluded that red wine induces more prevalent vasodilation than ethanol itself, and the polyphenols or tannic acid that it contains are likely responsible for this effect [55]. Notably, red wine appears to be more advantageous when compared to white wine, possibly due to the higher plasma polyphenols induced by its consumption [56].

Tea

Tea-derived polyphenols have also attracted interest during the last few decades as novel agents that can promote CV health [57–59]. With different types of tea consumed around the world, the question is raised as to which type offers the greatest benefit. In experimental studies, both green and black teas were associated with enhancement of eNOS activity and phosphorylation in bovine aortic endothelial cells, and induced vasorelaxation in rat aortic rings. Similarly, in human studies, both tea preparations induce the same increase in endothelial function in healthy subjects [60,61], suggesting that both green and black tea confer significant cardioprotective properties.

Overall Dietary Patterns

Mediterranean Diet

Recent research has focused on whole dietary patterns in order to establish the potential of overall nutrient intake to reduce CV risk. Much attention has focused on the traditional Mediterranean diet—a dietary pattern consisting of substantial quantities of fruits, vegetables, whole grains, lean meats (fish and poultry), olive oil, and regular (but moderate) red wine intake. The importance of this dietary pattern has been highlighted since the mid-1990s, when a prospective cohort study carried out in seven countries around the world noted that although CV mortality increased linearly with serum cholesterol levels in all countries, individuals in northern European countries were three times more likely to die than their southern Mediterranean counterparts at the same serum cholesterol level (~5.45 mmol/L)—suggesting that the dietary pattern of these countries afforded additional protection above and beyond traditional risk factors [62]. This hypothesis was recently confirmed by both a recent meta-analysis of prospective cohort studies [63] and a large-scale randomized control trial in which almost 7500 middle-aged volunteers (half female) were assigned to one of three groups—a Mediterranean diet supplemented with extra virgin olive oil, a Mediterranean diet supplemented with mixed nuts, or a control group advised to reduce dietary fat intake. Such was the success of both Mediterranean diet interventions that the study was concluded early after a median follow-up of 4.8 years following a 30% reduction in major adverse cardiac events in both the olive oil– and nut-supplemented diets groups when compared to controls [64]. Although the exact mechanisms for this cardioprotection are still not fully understood, one of the advantageous characteristics of the Mediterranean diet may be a high concentration of monounsaturated fats—such as those found in extra virgin olive oil. It is now indisputable that olive oil, being the cornerstone of this dietary pattern, has a variety of CV-advantageous properties, including lipid profile modification, blood pressure regulation, endothelial function enhancement, and oxidative stress reduction [9].

Western Diet

In contrast to the cardioprotective effects of the Mediterranean diet, the increasing global prevalence of a western-style diet consisting of a high meat, fat, and refined-carbohydrate intake alongside a low fruit and vegetable intake appears to be detrimental to CV health. In a prospective cohort study of 45,000 middle-aged men with no history of CVD, researchers identified a 30% risk reduction in individuals following the Mediterranean intervention and a 64% increase in risk for those following a western diet [65]. A later case–control study as part of the worldwide INTERHEART study confirmed and expanded upon these findings in a cohort from 52 countries, obtaining findings remarkably similar to those of the previous trial, with relative risk of acute myocardial infarction reported as 0.7 for those following the Mediterranean-style diet and 1.67 for those following the western-style one [66].

PHYSICAL ACTIVITY/FITNESS AND CV RISK

Alongside an increasing prevalence of unhealthy western-style dietary patterns, worldwide socioeconomic developments over the past 50 years have also seen rapid declines in levels of physical activity alongside a corresponding increase in sedentariness. The result of this global physical inactivity pandemic is that 31% of adults and 80% of adolescents currently fail to meet recommended daily exercise guidelines [67], leading physical inactivity to be responsible for the cause of up to 3.2 million deaths and loss of 69.3 million disability-adjusted life years (DALYs) per year [1]. The evidence for the ability of physical activity to decrease CVD risk is unequivocal. In a recent meta-analysis of 33 cohort studies involving over 880,000 participants with a follow-up time of 4–20 years, individuals reporting the highest levels of physical activity had a 33% reduced risk of all-cause mortality and a 35% reduced risk of CV mortality when compared with inactive individuals [68]. While this association is based entirely on observational studies, and as such cannot provide causality, the strength and consistency of the reported association provide compelling evidence for a cause-and-effect relationship. In addition, longitudinal studies assessing changes in physical activity and fitness have shown a decreased CVD risk in men who improve their physical fitness when compared to those who remain unfit [69]. The beneficial effect of physical activity is not restricted to adults, and recent work in children has shown that even at a young age, an inverse relationship exists between physical activity levels and the clustering of six CV disease risk factors (systolic blood pressure, triglyceride levels, total cholesterol/HDL ratio, insulin resistance, skinfold thickness, and cardiorespiratory fitness) [70]. While the vast majority of studies to date have assessed aerobic-based physical activity carried out in an individual's leisure time, both structured resistance training and occupational/commuting activity also appear protective against the development of CVD. Resistance training may benefit CVD risk through its positive effects on factors such as increased muscle mass, reduced hypertension, improved lipid profiles, decreased adiposity, and improved insulin resistance. A recent meta-analysis of physical activity while commuting has indicated an 11% reduction in overall CVD mortality in those who walk or cycle to work [71], highlighting the importance of non–leisure time physical activity in preventing CVD risk.

Recent focus has shifted to sedentariness as an independent risk factor for CV disease, as it is possible that an individual may meet recommended exercise guidelines of 30 min of moderate-intensity aerobic activity on 5 days of the week through a structured exercise plan but spend the rest of their waking day in a sedentary state (eg, working at a computer and watching TV). Recent pooled risk analyses assessing the impact of sitting and television viewing on life expectancy in the United States have estimated that up to 2 years could be added to population life expectancy if excessive sitting was reduced to less than 3 hours/day [72].

Similar to associations with physical activity, studies investigating the association between physical fitness (assessed via maximal exercise testing) and risk of CVD report even stronger relationships than those found using physical activity [73–75], possibly due to the overreporting of physical activity levels and recall bias when using subjective techniques such as questionnaires. While there is clearly a substantial reduction in CV risk in the most active members of the population compared to the least active ones, the levels of activity or fitness required to acquire this level of protection are most likely in excess of what many people are either able—or willing—to achieve, leading to the question of whether smaller increases in activity levels for inactive individuals would be sufficient to confer a CV benefit. Despite difficulties in standardizing the concept of a "dose" of exercise due to differences in both data collection and classification, evidence strongly suggests that progressive increases in physical activity or fitness result in progressive reductions in CV mortality risk. More importantly, this relationship appears to be curvilinear—with the greatest relative benefit occurring in individuals moving from little or no activity to moderate levels [76]. While these findings have previously been based on qualitative estimates of physical activity levels (such as low, moderate, and high), a recent meta-analysis has attempted to quantify this dose–response relationship for the first time [77], concluding that the risk of CHD is decreased by 14% in individuals meeting minimum recommended physical activity guidelines of 150 min/week of moderate-intensity aerobic exercise and that a further 6% decrease in risk can be achieved by doubling activity time to 300 min/week. Although further benefits continue to be observed with increasing levels of activity, these gains are relatively modest, supporting the notion that despite high physical activity levels being most beneficial, some activity is better than none at all. Although the mechanisms behind these benefits are still not fully understood, a wealth of research has shown the ability of physical exercise to exert a positive impact upon virtually every known risk factor for the development of CVD. These effects are due not only to improvements in traditional risk factors such as increased insulin sensitivity, lowered risk of type 2 diabetes, decreased obesity, increased cardiorespiratory fitness, and improved lipid and metabolic profiles [78], but also through direct effects on the vasculature itself—resulting in improvements in factors such as endothelial function, decreased inflammation and oxidative stress, decreased arterial stiffness, and increased angiogenesis (Fig. 5.2) [79].

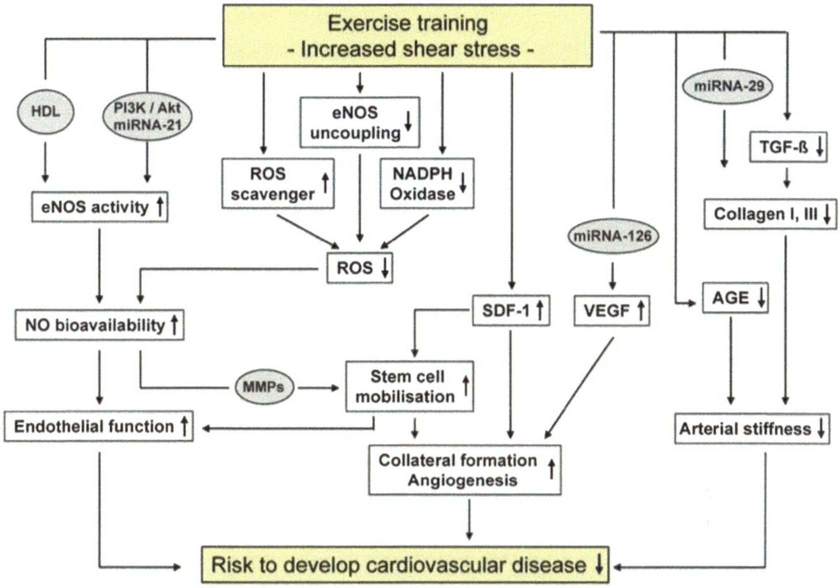

FIGURE 5.2 Possible mechanistic pathways by which increased levels of physical activity or exercise training may affect the vasculature and reduce the risk of CVD. *eNOS*, Endothelial nitric oxide synthase; *ROS*, reactive oxygen species; *SDF-1*, stromal cell-derived factor 1; *VEGF*, vascular endothelial growth factor; *TGF-β*, transforming growth factor beta; *AGE*, advanced glycation end products. *(From Schuler G, Adams V, Goto Y. Role of exercise in the prevention of cardiovascular disease: results, mechanisms, and new perspectives. Eur Heart J 2013;34:1790–9.)*

OBESITY AND CV RISK

The combination of Western-style dietary patterns and decreased levels of physical activity documented in this chapter has resulted in the recent emergence of a worldwide obesity epidemic. Although body mass index (BMI) is commonly used to describe excess body fat, and has been shown in large pooled risk analyses to strongly correlate with mortality [80], it is now clear that body fat distribution is more informative regarding individual CV risk. Excess visceral fat appears to be particularly detrimental to CV health, and it is associated with a cluster of metabolic abnormalities such as hyperinsulinemia, insulin resistance, diabetic dyslipidemia, hypertension, and pro-inflammatory and prothrombotic states. The presence of excess visceral fat, but not general adiposity, has also been shown to be independently associated with incident prediabetes and type 2 diabetes in obese subjects, providing a possible explanation for the heterogeneity in diabetes—and therefore CVD—risk observed in this population [81]. Simple measurements of waist circumference and waist/hip ratio have been shown to be an effective index of assessing underlying visceral adiposity, and as such appear to be more predictive of CVD than traditional BMI measures [82]. In addition to metabolic changes, increasing levels of adiposity also result in hemodynamic changes that act to increase cardiac work and result in concomitant negative effects on ventricular structure and systolic and diastolic ventricular function [83]. The mechanisms that link adiposity with increased CVD risk are yet to be clarified, as obesity not only modifies known CV risk factors such as diabetes and hypertension but also exerts an independent effect on CV health through the presence of excess adipose tissue itself. While adipose tissue was once considered a passive depot for storing excess calories, it is now recognized as a biologically active endocrine organ that has the ability to synthesize and secrete numerous regulatory molecules. These messengers include adiponectin, resistin, leptin, plasminogen activator inhibitor 1 (PAI-1), tumor necrosis factor-alpha (TNF-α), and interleukin 6 (IL-6)—all of which can upset endothelial function and promote atherogenesis [84]. Short-term lifestyle modifications resulting in weight reduction can positively affect this risk through improvements in metabolic profile, markers of cellular adhesion, and endothelial function. These modifications can be effective at any stage in life and result in sustained reductions in CV risk factors even if weight loss is not maintained long term [85]. The negative energy balance required for weight reduction can be achieved through a decrease in dietary energy intake, an increase in physical activity energy output, or a combination of the two. Unhealthy dietary patterns appear to be responsible for most, if not all, of the risk of obesity, and as such offer the primary method of intervention for achieving long-term weight gain. Despite short-term improvements in CV risk factors following weight loss in both observational and interventional studies [86], high attrition rates in the majority of studies make the assessment of long-term weight reduction and its impact on CVD mortality challenging. While a number of large-scale observational studies display inverse relationships between physical activity levels and the presence of overweight or obesity [87,88], results from interventional trials have indicated that physical activity levels well in excess of recommended daily guidelines may be required to attenuate weight gain—and may have even more modest effects on weight loss [89]. Numerous studies, however, have shown the effectiveness of physical activity to substantially decrease CV risk in obese individuals even in the absence of changes in body weight or BMI. Indeed, obese individuals with high cardiorespiratory fitness levels have been shown to be at lower risk of CVD mortality than their lean but unfit counterparts [90]. These changes are likely associated with a number of positive effects resulting from increased activity levels, including increased cardiorespiratory fitness, decreased waist circumference, decreased visceral adiposity, and improved metabolic profile [91].

Although weight reduction has been shown to reduce CV risk in most cases, an obesity paradox has been reported in certain conditions (eg, heart failure and dementia) whereby overweight and obese patients have a better prognosis than do their leaner counterparts [92]. The reasons for this are poorly understood but may result from reverse causality, with leaner individuals more likely to have suffered weight loss due to undernutrition or other underlying health issues.

SMOKING AND CV RISK

With the exception of aging, smoking is the single-most important risk factor for the development of CV disease, resulting in an estimated 6.3 million deaths annually [3]. While the vast majority of these deaths are attributable to active smoking, one in 10 occurs as a result of secondhand smoke. Smoking results in the loss of at least one decade of life expectancy and increases the risk of death in middle-aged men and women up to fivefold [93,94]. In a recent study analyzing trends in smoking-related mortality over the past 50 years in the United States, the relative risk of death from CHD was 2.50 and 2.86 for men and women, respectively [95]. This increased risk of CHD and stroke in women smokers compared to men is a consistent finding, and may result from an interaction with oral contraceptives acting in a synergistic fashion to increase risk. Similarly, substantial evidence also indicates that smoking acts synergistically with multiple other CVD risk factors to elevate risk above and beyond that caused by tobacco use itself, with smokers having an additional 15% increase in risk of hemorrhagic stroke for every 10 mmHg rise in systolic blood pressure when compared to their nonsmoking counterparts [96].

In addition to acting in tandem with other established risk factors, smoking increases CVD risk through multiple independent mechanisms. Nicotine contained within cigarettes is a sympathomimetic molecule that acutely elevates heart rate, cardiac contractility, and peripheral vasoconstriction, leading to repeated acute elevations in systemic blood pressure. With nicotine replacement therapy showing no adverse effects in relation to CVD events, however, it would appear that the risk posed by cigarettes may be predominantly attributable to other chemicals within the smoke [97]. In addition to acute blood pressure elevation, increases in arterial stiffness and pulse wave reflection may also act to further elevate central blood pressure even in the absence of any increases in brachial pressure, potentially increasing the risk of left ventricular hypertrophy and other organ damage [98]. Smoking promotes atherosclerotic plaque formation through a complex interaction between endothelial dysfunction, LDL-C oxidation, chronic inflammation, and platelet activation (Fig. 5.3) [99–103]. Endothelial function is decreased following exposure to tobacco smoke—an effect most likely mediated through decreased NO bioavailability as a result of increased levels of reactive oxygen species and free radicals. With the loss of the endothelium's antiadhesive properties conferred by NO, platelets and leukocytes adhere more readily to the arterial wall—a process helped by a simultaneous increase in platelet adhesion molecules such as platelet activating factor (PAF), thromboxane A2, von Willebrand factor, and PAI-1. Alongside endothelial dysfunction, smoking also results in a chronic inflammatory state characterized by the increased release of common inflammatory markers such as c-reactive protein (CRP), TNF-α, IL-6, and monocyte chemoattractant protein 1 (MCP-1) [104].

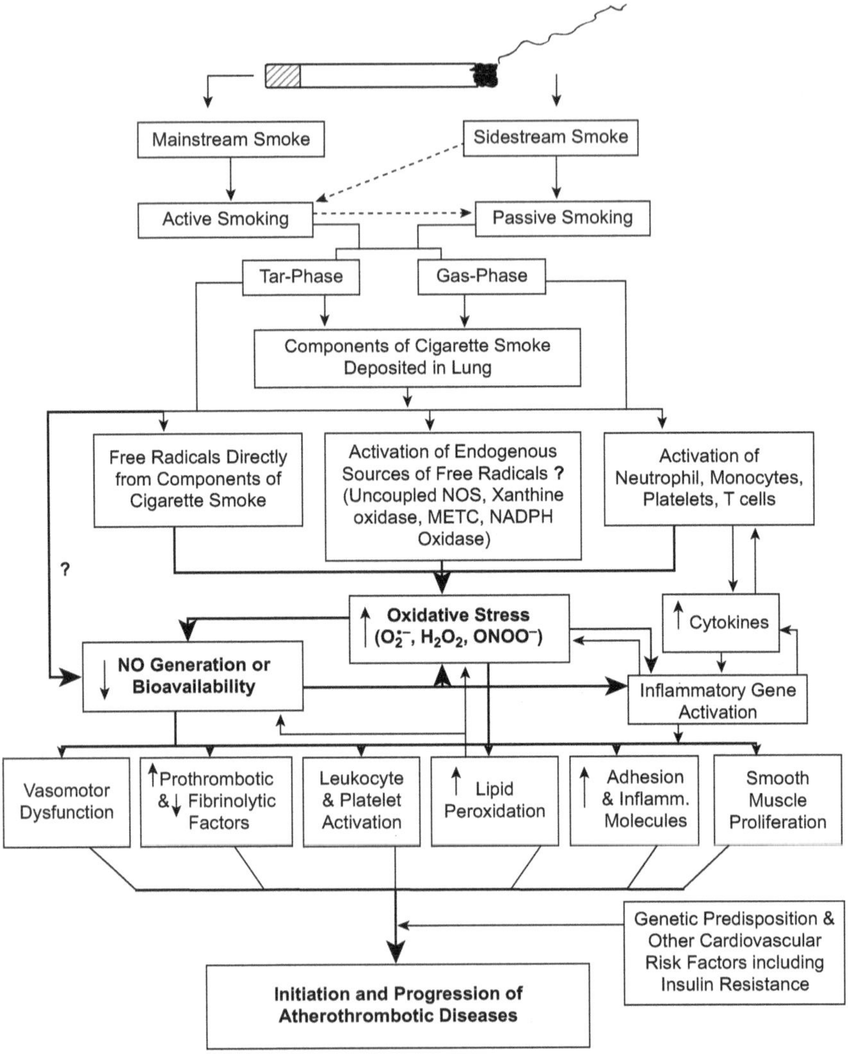

FIGURE 5.3 Potential pathways and mechanisms for cigarette-smoking-mediated cardiovascular dysfunction. The bold boxes and arrows represent the probable central mechanisms in the complex pathophysiology of cigarette-smoking-mediated atherothrombotic disease. H_2O_2, Hydrogen peroxide; *METC*, mitochondrial electron transport chain; *NADPH*, nicotinamide adenine dinucleotide phosphate reduced form; *NOS*, nitric oxide synthase; *OONO–*, peroxinitrite; O_2^-, superoxide. *(From Ambrose JA, Barua RS. The pathophysiology of cigarette smoking and cardiovascular disease: an update. J Am Coll Cardiol 2004;43:1731–7.)*

The association between smoking and CV disease risk follows a dose–response relationship. The upper limit of this relationship is not entirely clear, however; older and more contemporary studies showed, respectively, that the CV risk increases to a limit of 25 and 35–40 cigarettes per day [105]. While cigarettes are the most common form of smoking, all types of tobacco—including pipes, cigars, and filter cigarettes—are harmful. Research into the safety of e-cigarettes is currently lacking, but can be expected to rapidly expand following the recent boom in their popularity. Smoking cessation has well-established cardioprotective properties. A prospective double-blind randomized placebo-controlled clinical trial assessing the effects of five smoking cessation pharmacotherapies in 1504 smokers reported that after 1 year, endothelial function measured by FMD improved only in the group who quit smoking [106], while a beneficial effect on coronary vasomotor response has also been noted after only 1 month of cessation of smoking [107]. The functional properties of endothelial progenitor cells (EPCs) that play a fundamental role in angiogenesis and vascular repair can also be modified by a short-term cessation of smoking. Indeed, a 5-week absence of smoking was reported to induce alterations in numerous properties of EPCs as well as reductions in inflammation, oxidative stress, and plasma levels of asymmetric dimethyl-L-arginine [108].

PSYCHOLOGICAL FACTORS AND CV RISK

The contribution of psychological factors to CVD risk is often underappreciated. Socioeconomic differences can alter psychology in a way that predisposes to CVD and deteriorates the prognosis and quality of life of patients living with the disease (Fig. 5.4) [109]. Patients with depression are more prone to the development of CVD, and patients with CVD are more likely to suffer from depression than the general population. Thus, depression evidently contributes to CVD risk and, concomitantly, deteriorates its outcome in a graded fashion: the more severe the depression, the higher the mortality and coronary event recurrence rate [110]. The same conclusion is derived when the effect of anger is addressed, with anger and hostility not only increasing the risk for CHD in the healthy population, but also worsening the prognosis in those with preexisting CHD [111]. Despite abundant existing evidence suggesting that negative mood is associated with CVD, the underlying pathogenesis of this effect is not fully understood [112]. Mood disturbances have been shown to actively impair vasodilation, therefore exacerbating CVD while in healthy adults; hostility and anger suppression can induce endothelial dysfunction [113,114]. The link between psychological changes and endothelial dysfunction has been confirmed by experimental studies demonstrating an attenuation of the endothelium-dependent dilation in states of unpredictable chronic mild stress and depression [115]. In experimental animal models, rats rendered anhedonic after 8 weeks of chronic mild stress presented with decreased endothelium-dependent vasodilation of their mesenteric arteries as a result of suppression of the endothelium-dependent hyperpolarization-like relaxation. This finding suggests that major depression alters peripheral resistance and organ perfusion [116]. High total trait anger has recently been shown to be a potential risk factor for incident type 2 diabetes [117], providing a novel mechanism by which this condition may contribute to total CVD risk. Even the subjective feeling of exhaustion seems to play a decisive role, with epidemiological evidence showing a link between vital

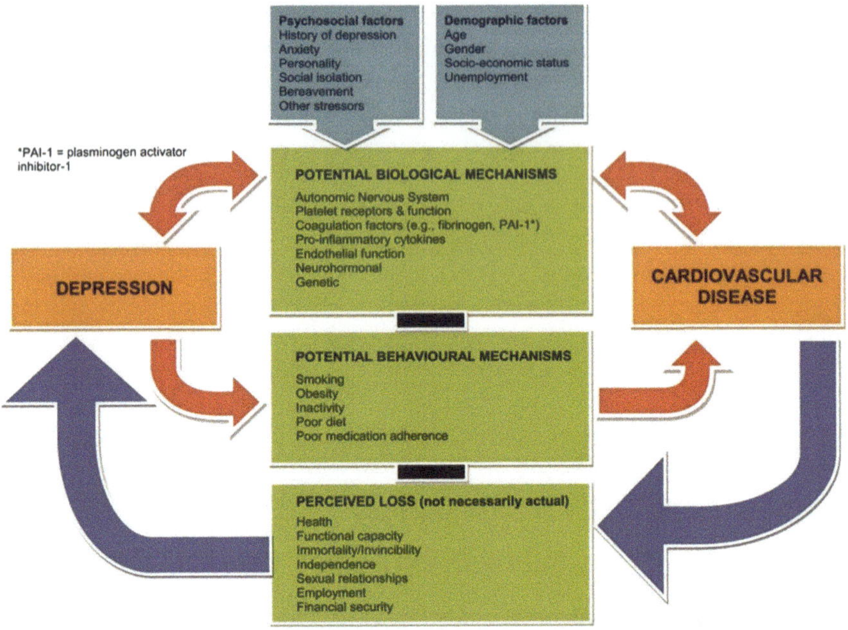

FIGURE 5.4 Potential factors linking depression and CVD. Biological and psychological changes caused by numerous negative psychosocial and demographic factors may not only predispose individuals with depression to increased CVD risk, but conversely predispose those with CVD to an increased risk of depression. *(From Hare DL, Toukhsati SR, Johansson P, Jaarsma T. Depression and cardiovascular disease: a clinical review. Eur Heart J 2014;35:1365–72.)*

exhaustion and cerebrocardiovascular disease, while fatigue appears to hinder normal endothelial function. In healthy people, endothelial function, assessed by FMD before and after work, has been shown to correlate with the subjective estimation of an individual's fatigue [118]. In the quest for a pathophysiological mechanism that underlies the observed link between negative mood and CVD, changes in the autonomic nervous system and its interactions with endothelial function seem the most likely explanation. In the constant endeavor to maintain vascular-tone homeostasis, the sympathetic nerves and the vascular endothelial cells play antagonistic roles. Thus, any harmful event on either the autonomic nervous system or endothelial function may negatively influence the existing balance and contribute to disease progression [119]. Studies using various techniques and behavioral therapies with the aim to reduce psychological stress (eg, yoga or laughter generation) display a positive influence on endothelial function [120–122], while a large meta-analysis of prospective, observational, and cohort studies has concluded that psychological wellbeing (ie, positive mood, happiness, energy, life satisfaction, hopefulness, optimism, and humor) was inversely linked to CV mortality both in healthy populations and in patients with medical comorbidities. Apart from improvements in endothelial function, other mechanisms involving the autonomic system, neuroendocrine activity, inflammatory pathways, and platelet activation are also likely to contribute to this reported benefit.

CONCLUSION

Despite the presence of nonmodifiable risk factors for CVD such as aging and gender, numerous lifestyle behaviors have been identified that can significantly increase the risk of developing the disease. While these behaviors are numerous and varied, they share a common trait in that they are modifiable, and existing evidence suggests that promoting these modifications has the potential to significantly reduce CVD burden. Interactions between an individual's genetic heritage and these environmental exposures, however, influence CV health in a way that is still largely unpredictable. Further studies are required in order to identify people most at risk of CVD due to "unhealthy" lifestyle choices and establish who will optimally benefit from targeted lifestyle or pharmacological interventions.

REFERENCES

[1] World Health Organization. Global status report on noncommunicable diseases 2014. 2014.

[2] Yusuf, S. et al. Effect of potentially modifiable risk factors associated with myocardial infarction in 52 countries (the INTERHEART study): case-control study. Lancet 364, 937–52.

[3] Murray CJL, et al. Disability-adjusted life years (DALYs) for 291 diseases and injuries in 21 regions, 1990–2010: a systematic analysis for the Global Burden of Disease Study 2010. Lancet 2012;380:2197–223.

[4] Mozaffarian D, Katan MB, Ascherio A, Stampfer MJ, Willett WC. Trans fatty acids and cardiovascular disease. N Engl J Med 2006;354:1601–13.

[5] KEYS A. Coronary heart disease in seven countries. Circulation 1970;41:186–95.

[6] Skeaff CM, Miller J. Dietary fat and coronary heart disease: summary of evidence from prospective cohort and randomised controlled trials. Ann Nutr Metab 2009;55:173–201.

[7] Siri-tarino PW, Sun Q, Hu FB, Krauss RM. Meta-analysis of prospective cohort studies evaluating the association of saturated fat with cardiovascular disease 1–5. Am J Clin Nutr 2010;91:535–46.

[8] Astrup A, et al. The role of reducing intakes of saturated fat in the prevention of cardiovascular disease: where does the evidence stand in 2010? Am J Clin Nutr 2011;93:684–8.

[9] López-Miranda J, et al. Olive oil and health: summary of the II international conference on olive oil and health consensus report, Jaén and Córdoba (Spain) 2008. Nutr Metab Cardiovasc Dis 2010;20:284–94.

[10] Jakobsen MU, et al. Major types of dietary fat and risk of coronary heart disease: a pooled analysis of 11 cohort studies 1–3. Am J Clin Nutr 2009;89:1425–33.

[11] Ludwig DS. Dietary fiber, weight gain, and cardiovascular disease risk factors in young adults. JAMA 1999;282:1539.

[12] McKeown NM, Meigs JB, Liu S, Wilson PW, Jacques PF. Whole-grain intake is favorably associated with metabolic risk factors for type 2 diabetes and cardiovascular disease in the Framingham Offspring Study. Am J Clin Nutr 2002;76:390–8.

[13] Siri-tarino PW, Sun Q, Hu FB, Krauss RM. Saturated fat, carbohydrate, and cardiovascular disease. Am J Clin Nutr 2010:502–9.

[14] Jakobsen MU, et al. Intake of carbohydrates compared with intake of saturated fatty acids and risk of myocardial infarction: importance of the glycemic index. Am J Clin Nutr 2010; 91:1764–8.

[15] Aaron KJ, Sanders PW. Role of dietary salt and potassium intake in cardiovascular health and disease: a review of the evidence. Mayo Clin Proc 2013;88:987–95.

[16] Kalogeropoulos AP, et al. Dietary sodium content, mortality, and risk for cardiovascular events in older adults: the Health, Aging, and Body Composition (Health ABC) Study. JAMA Intern Med 2015;175:410–9.

[17] Strazzullo P, D'Elia L, Kandala N-B, Cappuccio FP. Salt intake, stroke, and cardiovascular disease: meta-analysis of prospective studies. BMJ 2009;339:b4567.

[18] Bibbins-Domingo K, et al. Projected effect of dietary salt reductions on future cardiovascular disease. N Engl J Med 2010;362:590–9.

[19] Institute of Medicine (US) Committee on Strategies to Reduce Sodium Intake. Strategies to reduce sodium intake in the United States. In: Henney JE, Taylor CL, Boon CS, editors. 2010. at: http://www.ncbi.nlm.nih.gov/books/NBK50956/.

[20] O'Donnell MJ, et al. Urinary sodium and potassium excretion and risk of cardiovascular events. JAMA 2011;306:2229–38.

[21] Pan A, et al. Red meat consumption and mortality: results from 2 prospective cohort studies. Arch Intern Med 2012;172:555–63.

[22] Micha R, Michas G, Mozaffarian D. Unprocessed red and processed meats and risk of coronary artery disease and type 2 diabetes–an updated review of the evidence. Curr Atheroscler Rep 2012;14:515–24.

[23] Koeth RA, et al. Intestinal microbiota metabolism of L-carnitine, a nutrient in red meat, promotes atherosclerosis. Nat Med 2013;19:576–85.

[24] Wang Z, et al. Gut flora metabolism of phosphatidylcholine promotes cardiovascular disease. Nature 2011;472:57–63.

[25] He K, et al. Fish consumption and risk of stroke in men. JAMA 2002;288:3130–6.

[26] He K, et al. Accumulated evidence on fish consumption and coronary heart disease mortality: a meta-analysis of cohort studies. Circulation 2004;109:2705–11.

[27] Oomen CM, et al. Fish consumption and coronary heart disease mortality in Finland, Italy, and The Netherlands. Am J Epidemiol 2000;151:999–1006.

[28] Schini-Kerth VB, Etienne-Selloum N, Chataigneau T, Auger C. Vascular protection by natural product-derived polyphenols: in vitro and in vivo evidence. Planta Med 2011;77:1161–7.

[29] Stoclet J-C, et al. Vascular protection by dietary polyphenols. Eur J Pharmacol 2004;500: 299–313.

[30] Li S-H, Tian H-B, Zhao H-J, Chen L-H, Cui L-Q. The acute effects of grape polyphenols supplementation on endothelial function in adults: meta-analyses of controlled trials. PLoS One 2013;8:e69818.

[31] Rimbach G, Egert S, de Pascual-Teresa S. Chocolate: (un)healthy source of polyphenols? Genes Nutr 2011;6:1–3.

[32] Vlachopoulos C, et al. Effect of dark chocolate on arterial function in healthy individuals. Am J Hypertens 2005;18:785–91.

[33] Ried K, Sullivan T, Fakler P, Frank OR, Stocks NP. Does chocolate reduce blood pressure? A meta-analysis. BMC Med 2010;8:39.

[34] Faridi Z, Njike VY, Dutta S, Ali A, Katz DL. Acute dark chocolate and cocoa ingestion and endothelial function: a randomized controlled crossover trial. Am J Clin Nutr 2008;88:58–63.

[35] Shiina Y, et al. Acute effect of oral flavonoid-rich dark chocolate intake on coronary circulation, as compared with non-flavonoid white chocolate, by transthoracic Doppler echocardiography in healthy adults. Int J Cardiol 2009;131:424–9.

[36] Maersk M, et al. Satiety scores and satiety hormone response after sucrose-sweetened soft drink compared with isocaloric semi-skimmed milk and with non-caloric soft drink: a controlled trial. Eur J Clin Nutr 2012;66:523–9.

[37] Malik VS, Pan A, Willett WC, Hu FB. Sugar-sweetened beverages and weight gain in children and adults: a systematic review and meta-analysis. Am J Clin Nutr 2013;98:1084–102.

[38] Mozaffarian D, Hao T, Rimm EB, Willett WC, Hu FB. Changes in diet and lifestyle and long-term weight gain in women and men. N Engl J Med 2011;364:2392–404.

[39] Chen L, et al. Reduction in consumption of sugar-sweetened beverages is associated with weight loss: the PREMIER trial. Am J Clin Nutr 2009;89:1299–306.

[40] Imamura F, et al. Consumption of sugar sweetened beverages, artificially sweetened beverages, and fruit juice and incidence of type 2 diabetes: systematic review, meta-analysis, and estimation of population attributable fraction. BMJ 2015;351:h3576.

[41] Renaud S, de Lorgeril M. Wine, alcohol, platelets, and the French paradox for coronary heart disease. Lancet 1992;339:1523–6.

[42] Marmot MG, Shipley MJ, Rose G, Thomas B. Alcohol and mortality: a U-shaped curve. Lancet 1981;317:580–3.

[43] Kozararevic D. Frequency of alcohol consumption and morbidity and mortality: the Yugoslavia cardiovascular disease study. Lancet 1980;315:613–6.

[44] Shaper AG, Wannamethee G, Walker M. Alcohol and mortality in British men: explaining the U-shaped curve. Lancet 1988;332:1267–73.

[45] Wannamethee S, Shaper A. Lifelong teetotallers, ex-drinkers and drinkers: mortality and the incidence of major coronary heart disease events in middle-aged British men. Int J Epidemiol 1997;26:523–31.

[46] Klatsky AL. Wine, liquor, beer, and mortality. Am J Epidemiol 2003;158:585–95.

[47] Suzuki K, et al. Moderate alcohol consumption is associated with better endothelial function: a cross sectional study. BMC Cardiovasc Disord 2009;9:8.

[48] Lippi G, Franchini M, Favaloro EJ, Targher G. Moderate red wine consumption and cardiovascular disease risk: beyond the 'French paradox'. Semin Thromb Hemost 2010;36:59–70.

[49] Frankel E, German J, Kinsella J, Parks E, Kanner J. Inhibition of oxidation of human low-density lipoprotein by phenolic substances in red wine. Lancet 1993;341:454–7.

[50] Freedman JE, et al. Select flavonoids and whole juice from purple grapes inhibit platelet function and enhance nitric oxide release. Circulation 2001;103:2792–8.

[51] Pendurthi UR, Williams JT, Rao LVM. Resveratrol, a polyphenolic compound found in wine, inhibits tissue factor expression in vascular cells: a possible mechanism for the cardiovascular benefits associated with moderate consumption of wine. Arterioscler Thromb Vasc Biol 1999;19:419–26.

[52] Wollny T, et al. Modulation of haemostatic function and prevention of experimental thrombosis by red wine in rats: a role for increased nitric oxide production. Br J Pharmacol 1999; 127:747–55.

[53] Micallef M, Lexis L, Lewandowski P. Red wine consumption increases antioxidant status and decreases oxidative stress in the circulation of both young and old humans. Nutr J 2007;6:27.

[54] Agewall S, et al. Does a glass of red wine improve endothelial function? Eur Heart J 2000; 21:74–8.

[55] Porteri E, et al. Vasodilator effects of red wines in subcutaneous small resistance artery of patients with essential hypertension. Am J Hypertens 2010;23:373–8.

[56] Pignatelli P, et al. Polyphenols synergistically inhibit oxidative stress in subjects given red and white wine. Atherosclerosis 2006;188:77–83.
[57] Riemersma RA. Tea flavonoids and cardiovascular health. QJM 2001;94:277–82.
[58] Stangl V, Dreger H, Stangl K, Lorenz M. Molecular targets of tea polyphenols in the cardiovascular system. Cardiovasc Res 2007;73:348–58.
[59] Ras RT, Zock PL, Draijer R. Tea consumption enhances endothelial-dependent vasodilation; a meta-analysis. PLoS One 2011;6:e16974.
[60] Jochmann N, et al. The efficacy of black tea in ameliorating endothelial function is equivalent to that of green tea. Br J Nutr 2008;99:863–8.
[61] Lorenz M, et al. Green and black tea are equally potent stimuli of NO production and vasodilation: new insights into tea ingredients involved. Basic Res Cardiol 2009;104:100–10.
[62] Verschuren WM, et al. Serum total cholesterol and long-term coronary heart disease mortality in different cultures. Twenty-five-year follow-up of the seven countries study. JAMA 1995;274:131–6.
[63] Sofi F, Abbate R, Gensini GF, Casini A. Accruing evidence on benefits of adherence to the Mediterranean diet on health: an updated systematic review and meta-analysis. Am J Clin Nutr 2010;92:1189–96.
[64] Estruch R, et al. Primary prevention of cardiovascular disease with a Mediterranean diet. N Engl J Med 2013;368:1279.
[65] Hu FB, et al. Prospective study of major dietary patterns and risk of coronary heart disease in men. Am J Clin Nutr 2000;72:912–21.
[66] Iqbal R, et al. Dietary patterns and the risk of acute myocardial infarction in 52 countries: results of the INTERHEART study. Circulation 2008;118:1929–37.
[67] Hallal PC, et al. Global physical activity levels: surveillance progress, pitfalls, and prospects. Lancet 2012;380:247–57.
[68] Nocon M, Hiemann T, Muller-Riemenschneider F, Thalau F, Roll S, Willich SN. Association of physical activity with all-cause and cardiovascular mortality: a systematic review and meta-analysis. Eur J Cardio-Vasc Prev Rehabil 2008;15:239–46.
[69] Blair SN. Changes in physical fitness and all-cause mortality. JAMA 1995;273:1093.
[70] Andersen LB, et al. Physical activity and clustered cardiovascular risk in children: a cross-sectional study (The European Youth Heart Study). Lancet 2006;368:299–304.
[71] Hamer M, Chida Y. Active commuting and cardiovascular risk: a meta-analytic review. Prev Med 2008;46:9–13.
[72] Katzmarzyk PT, Lee I-M. Sedentary behaviour and life expectancy in the USA: a cause-deleted life table analysis. BMJ Open 2012;2:e000828.
[73] Blair SN, Cheng Y, Holder JS. Is physical activity or physical fitness more important in defining health benefits? Med Sci Sports Exerc 2001;33:S379–99.
[74] Lee C, Blair SN. Cardiorespiratory fitness and stroke mortality in men. Med Sci Sports Exerc 2002;34:592–5.
[75] Church TS, LaMonte MJ, Barlow CE, Blair SN. Cardiorespiratory fitness and body mass index as predictors of cardiovascular disease mortality among men with diabetes. Arch Intern Med 2005;165:2114–20.
[76] Kokkinos P, et al. Exercise capacity and mortality in black and white men. Circulation 2008;117:614–22.
[77] Sattelmair J, et al. Dose response between physical activity and risk of coronary heart disease: a meta-analysis. Circulation 2011;124:789–95.
[78] Gielen S, Schuler G, Adams V. Cardiovascular effects of exercise training: molecular mechanisms. Circulation 2010;122:1221–38.

[79] Schuler G, Adams V, Goto Y. Role of exercise in the prevention of cardiovascular disease: results, mechanisms, and new perspectives. Eur Heart J 2013;34:1790–9.

[80] Berrington de Gonzalez A, et al. Body-mass index and mortality among 1.46 million white adults. N Engl J Med 2010;363:2211–9.

[81] Neeland IJ, et al. Dysfunctional adiposity and the risk of prediabetes and type 2 diabetes in obese adults. JAMA 2012;308:1150–9.

[82] Yusuf S, et al. Obesity and the risk of myocardial infarction in 27,000 participants from 52 countries: a case-control study. Lancet 2005;366:1640–9.

[83] Alpert MA. Obesity cardiomyopathy: pathophysiology and evolution of the clinical syndrome. Am J Med Sci 2001;321:225–36.

[84] Kershaw EE, Flier JS. Adipose tissue as an endocrine organ. J Clin Endocrinol Metab 2004;89:2548–56.

[85] Charakida M, et al. Lifelong patterns of BMI and cardiovascular phenotype in individuals aged 60–64 years in the 1946 British birth cohort study: an epidemiological study. Lancet Diabetes Endocrinol 2014;2.

[86] Van Gaal LF, Wauters MA, De Leeuw IH. The beneficial effects of modest weight loss on cardiovascular risk factors. Int J Obes Relat Metab Disord 1997;21(Suppl. 1):S5–9.

[87] Martínez-González MA, Martínez JA, Hu FB, Gibney MJ, Kearney J. Physical inactivity, sedentary lifestyle and obesity in the European Union. Int J Obes Relat Metab Disord 1999;23:1192–201.

[88] King GA, et al. Relationship of leisure-time physical activity and occupational activity to the prevalence of obesity. Int J Obes Relat Metab Disord 2001;25:606–12.

[89] Donnelly JE, et al. American College of Sports Medicine Position Stand. Appropriate physical activity intervention strategies for weight loss and prevention of weight regain for adults. Med Sci Sports Exerc 2009;41:459–71.

[90] Lee CD, Blair SN, Jackson AS. Cardiorespiratory fitness, body composition, and all-cause and cardiovascular disease mortality in men. Am J Clin Nutr 1999;69:373–80.

[91] Ross R, Janiszewski PM. Is weight loss the optimal target for obesity-related cardiovascular disease risk reduction? Can J Cardiol 2008;24(Suppl. D):25D–31D.

[92] Lavie CJ, Milani RV, Ventura HO. Obesity and cardiovascular disease: risk factor, paradox, and impact of weight loss. J Am Coll Cardiol 2009;53:1925–32.

[93] CDC's Office on Smoking and Health, Smoking and Tobacco Use; Fact Sheet; Tobacco-Related Mortality, 2015. at:http://www.cdc.gov/tobacco/data_statistics/fact_sheets/health_effects/tobacco_related_mortality/.

[94] Pirie K, Peto R, Reeves GK, Green J, Beral V. The 21st century hazards of smoking and benefits of stopping: a prospective study of one million women in the UK. Lancet 2013; 381:133–41.

[95] Thun MJ, et al. 50-year trends in smoking-related mortality in the United States. N Engl J Med 2013;368:351–64.

[96] Nakamura K, et al. Cigarette smoking, systolic blood pressure, and cardiovascular diseases in the Asia-Pacific region. Stroke 2008;39:1694–702.

[97] Hubbard R, et al. Use of nicotine replacement therapy and the risk of acute myocardial infarction, stroke, and death. Tob Control 2005;14:416–21.

[98] Doonan RJ, et al. The effect of smoking on arterial stiffness. Hypertens Res 2010;33: 398–410.

[99] Celermajer DS, et al. Cigarette smoking is associated with dose-related and potentially reversible impairment of endothelium-dependent dilation in healthy young adults. Circulation 1993;88:2149–55.

[100] Scheffler E, et al. Alteration of plasma low density lipoprotein from smokers. Atherosclerosis 1990;82:261–5.
[101] Pamukcu B, Oflaz H, Onur I, Cimen A, Nisanci Y. Effect of cigarette smoking on platelet aggregation. Clin Appl Thromb Hemost 2011;17:E175–80.
[102] Messner B, Bernhard D. Smoking and cardiovascular disease: mechanisms of endothelial dysfunction and early atherogenesis. Arterioscler Thromb Vasc Biol 2014;34:509–15.
[103] Ambrose JA, Barua RS. The pathophysiology of cigarette smoking and cardiovascular disease: an update. J Am Coll Cardiol 2004;43:1731–7.
[104] Rom O, Avezov K, Aizenbud D, Reznick AZ. Cigarette smoking and inflammation revisited. Respir Physiol Neurobiol 2013;187:5–10.
[105] U.S. Department of Health and Human Services, C. for D. C. and P., National Center for Chronic Disease Prevention and Health Promotion, O. on S. & Health, A. Surgeon General's report: the health consequences of Smoking—50 years of progress. 2014.
[106] Johnson HM, et al. Effects of smoking and smoking cessation on endothelial function: 1-year outcomes from a randomized clinical trial. J Am Coll Cardiol 2010;55:1988–95.
[107] Morita K, et al. Smoking cessation normalizes coronary endothelial vasomotor response assessed with 15O-water and PET in healthy young smokers. J Nucl Med 2006;47:1914–20.
[108] Puls M, et al. Effect of smoking cessation on the number and adhesive properties of early outgrowth endothelial progenitor cells. Int J Cardiol 2011;152:61–9.
[109] Skodova Z, et al. Socioeconomic differences in psychosocial factors contributing to coronary heart disease: a review. J Clin Psychol Med Settings 2008;15:204–13.
[110] Hare DL, Toukhsati SR, Johansson P, Jaarsma T. Depression and cardiovascular disease: a clinical review. Eur Heart J 2014;35:1365–72.
[111] Chida Y, Steptoe A. The association of anger and hostility with future coronary heart disease: a meta-analytic review of prospective evidence. J Am Coll Cardiol 2009;53:936–46.
[112] Cooper DC, et al. Adverse impact of mood on flow-mediated dilation. Psychosom Med 2010;72:122–7.
[113] Schott LL, Kamarck TW, Matthews KA, Brockwell SE, Sutton-Tyrrell K. Is brachial artery flow-mediated dilation associated with negative affect? Int J Behav Med 2009;16:241–7.
[114] Osika W, et al. Anger, depression and anxiety associated with endothelial function in childhood and adolescence. Arch Dis Child 2011;96:38–43.
[115] d'Audiffret AC, et al. Depressive behavior and vascular dysfunction: a link between clinical depression and vascular disease? J Appl Physiol 2010;108:1041–51.
[116] Bouzinova EV, et al. Association between endothelial dysfunction and depression-like symptoms in chronic mild stress model of depression. Psychosom Med 2014;76:268–76.
[117] Abraham S, et al. Trait anger but not anxiety predicts incident type 2 diabetes: the Multi-Ethnic Study of Atherosclerosis (MESA). Psychoneuroendocrinology 2015;60:105–13.
[118] Ohno Y, et al. The diagnostic value of endothelial function as a potential sensor of fatigue in health. Vasc Health Risk Manag 2010;6:135–44.
[119] Harris KF, Matthews KA. Interactions between autonomic nervous system activity and endothelial function: a model for the development of cardiovascular disease. Psychosom Med 2004;66:153–64.
[120] Miller M, et al. Impact of cinematic viewing on endothelial function. Heart 2006;92:261–2.
[121] Wang J-S, Lan C, Chen S-Y, Wong M-K. Tai Chi Chuan training is associated with enhanced endothelium-dependent dilation in skin vasculature of healthy older men. J Am Geriatr Soc 2002;50:1024–30.
[122] Sivasankaran S, et al. The effect of a six-week program of yoga and meditation on brachial artery reactivity: do psychosocial interventions affect vascular tone? Clin Cardiol 2006;29:393–8.

Chapter 6

The Contribution of Pollutants and Environmental Chemicals in Cardiovascular Disease

G. Hatzis[1,2], K. Karatolios[2]
[1]Athens University Medical School, Athens, Greece; [2]Philipps University Marburg, Germany

INTRODUCTION

There is no doubt that cardiovascular disease is the leading cause of mortality in Western societies, being simultaneously a big socioeconomic problem and a heavy burden for all health systems worldwide. Cardiovascular disease is the net result of complex mechanisms, which are based on genetic as well as environmental processes. Emerging data and comprehensive studies throughout the past three decades have clearly shown that environmental pollution plays a key role in the underlying processes that lead to and favor atherosclerosis, the main pathological substrate of cardiovascular disease. The term *pollution* defines the introduction of harmful substances or products into the environment; these substances are either natural or secondary products coming from human activities. These products are called pollutants. A plethora of pollutants have been connected with the appearance and evolution of cardiovascular disease. A detailed presentation of all these factors exceeds the ability of analysis in a defined framework. The scope of this chapter is to refer to the main traditional environmental pollutants that contribute to the mechanisms of atherosclerosis. These pollutants are mainly divided into air and metal pollutants. The main representatives of air pollutants include particulate matter, ozone, and environmental smoke, while in the category of metals pollutants cadmium, mercury, lead, and arsenic are the main representatives (Table 6.1).

AIR POLLUTANTS

Over the last decade a growing body of epidemiological and clinical evidence has led to a heightened concern about the potential deleterious effects of ambient air pollution on health and its relation to all forms of cardiovascular disease. Air pollution is a mixture of natural (not made) substances in the

Cardiovascular Diseases.

TABLE 6.1 Main Environmental Pollutants Linked With Cardiovascular Disease

Air Pollutants	Main Metal Pollutants
Particulate matter $PM_{2.5}$ (μg/m^3)	Lead
Particulate matter PM_{10} (μg/m^3)	Cadmium
Ozone	Mercury
Environmental smoke	Arsenic
Carbon monoxide, nitrogen dioxide, sulfur dioxide	

air we breathe. Typically it is separated into indoor and outdoor pollution. Outdoor pollution includes fine particles produced by the burning of fusil fuels, ground-level ozone, noxious gases, and tobacco smoke. On the other hand, indoor pollution includes household products and chemicals, building materials, and tobacco smoke. The aim of this chapter is to summarize the existence and the action of the most important air pollutants on cardiovascular homeostasis as well as to refer to the basic mechanisms of their pathophysiological activities.

Particulate Matter and Cardiovascular Disease

Atmospheric particulate matter, also known as particulates, is microscopic liquid or matter suspended in the earth's atmosphere. The adverse effects of the particulate matter substances, especially according to their size, are of particular interest for governments and health organizations around the world [1]. The substances are divided broadly into the category containing the particles with a size smaller than 2.5 μm and those with a size up to 10 μm. The source of these substances comes from mainly two categories: natural phenomena, such as volcanoes, wildfires, or land dust; and human activities, such as carbon monoxide (CO) from vehicle exhausts or sulfur dioxide (SO_2) from industrial process, with the subclass coming from human activities being the most harmful for human health [2]. In the study from Gold et al., a significant increase of 0.5% to 1.5% was observed in incidence of cardiovascular disease for every increase of 5–6 μm/m^3 in the levels of $PM_{2.5}$. In the same study, it was shown that the acute increase of $PM_{2.5}$ resulted in greater cardiovascular mortality in comparison with the effects on the respiratory tract [3]. Furthermore, in a large study consisting of more than 500,000 participants with a long follow-up, the levels of $PM_{2.5}$ were proportionally correlated with the risk of ischemic heart disease, heart failure, arrhythmias, and cardiac arrest [4]. These data prompted the American Heart Association to issue an updated announcement concerning the role of the air pollutants in cardiovascular health in 2009. In the United States, a study in Boston showed a significant risk of myocardial infarction according to $PM_{2.5}$ levels [5], while in the Intermountain Heart Collaborative Study a 4.5% increase in coronary artery disease was associated with every 10 μg/m^3 increase in $PM_{2.5}$ [6]. Analogous data exist on the role of PM_{10} in cardiovascular health. The National Morbidity, Mortality, and Air Pollution Study showed a 0.68% increase in cardiopulmonary mortality for every 10 μg/m^3 increase in PM_{10} on the day before death [7–9]. The Air Pollution and Health European Approach (APHEA-2) study in Europe showed an increase of 0.76 in cardiovascular mortality for every 10 μg/m^3 increase in PM_{10} [8–10]. Importantly, the increase in the level of PM_{10} resulted in an increase in hospitalization for congestive heart failure as well as for ischemic heart disease [11]. Similar data exist on the role of this pollutant in the prevalence of stroke in the United States and in Europe [5,6,11–14]. The main mechanism that mediates all these activities is the induction of oxidative stress and all of the mechanisms that interfere with the cell behavior by acting at the DNA level [15].

Ozone Pollution

Ozone (O_3) is described as a colorless gas molecule consisting of three atoms of oxygen, highly reactive and with a high-oxidizing power. The locality of ozone production determines its action and its tendency to damage the human homeostasis or to promote its beneficial effects on humans. The sources of ozone precursor pollutants are both anthropic (motor vehicles, combustion processes, thermoelectric plants, chemical solvents) and natural (woods and forests emit highly reactive volatile organic compounds, like terpenes). To some extent, ozone is present in the troposphere because of the natural interrange with the stratosphere, and its concentration may vary between 20 and 80 μg/m^3. Higher concentrations of ozone are caused by photochemical smog. Interestingly, upper or under cutoff values have not been set for ozone concerning the impact to the human homeostasis, which means that apparently there is no threshold dose below which no effects occur. A plethora of studies have shown the direct and the indirect action of ozone on the vasculature. One potential mechanism is vasoconstriction, which is a major feature of cardiovascular disease [16]. Another mechanism is the induction of hypertension, which is an established risk factor for cardiovascular disease [17]. Experimental data also suggest that ozone precipitates to arrhythmias and to an increased heart rate, which in turn induces the presence and evolution of cardiovascular disease [18–21]. Other data suggest that ozone also induces the production of inflammatory markers, which in turn induce the sequence of events that lead to atherosclerosis [22–25].

In addition to the impact of ozone leading to pulmonary diseases, many epidemiological studies have also concentrated on studying the role of ozone on the cardiovascular system. A 10 ppb increase in daily ozone was associated with increases in daily mortality of 2.84% (95% CI, 0.95–4.77%) for four European cities [20], 0.61% (95% CI, 0.38–1.60%) for seven Spanish cities, 1.40% (95% CI, 0.68–2.12%) for six French cities [26], and 0.43% (95% CI, 0.23–0.63%) for 80 US urban centers from 1987 to 1994. Another large study from various European cities showed that a 10 mg/m^3 increase in the 1-h O^3 concentration was associated with a 0.45% increase in the number of cardiovascular deaths, while similar data come also from Canada [27,28]. Other data show an increase of acute cardiac events according to ozone concentration in Europe as well as in America, while the effects of ozone in the elderly seem to be more deleterious [29–31]. All these data underline the role of ozone in human health, but of course the exact mechanisms remain obscure and further studies are needed in order to understand the real effects of ozone on the vasculature.

Environmental Smoke

Smoking is an established major risk factor for cardiovascular disease and indisputable data suggest that smoking cessation leads to an improvement of all risk models regarding the prediction and the incidence of any form of cardiovascular disease [32,33]. (A detailed analysis of this fact is out of the scope of the present review.) Given the undisputable role of smoking in atherogenesis, it is not a surprise that in the last decades a plethora of scientists have focused on the role of secondhand smoking, or passive or environmental tobacco smoke, on cardiovascular health. A detailed presentation of all related studies is beyond the present review, however, references to some of the main studies as well as to the main pathophysiological mechanisms will be given. Current estimates suggest that passive smoking increases the risk of heart disease by 30%, accounting for at least 35,000 deaths in United States annually [34–38]. Analogous data exist on the role of passive smoking on the incidence of stroke and acute myocardial infarction [39–42]. Epidemiological data on the relationship between passive smoking and cardiovascular disease have been accumulated since the mid-1980s. In the meta-analysis of Glantz et al. based on 10 epidemiological studies, it was shown that passive smoking causes a 30% increase in the incidence of coronary heart disease as well as in the incidence of myocardial infarction [34]. Similar results came from a post hoc analysis from the same researchers few years later [35]. In another meta-analysis it was shown that the subjects that lived in a smoke-rich environment had a relative risk of 1.3 for developing ischemic heart disease compared to subjects that were not exposed to passive smoking [37]. In another meta-analysis the similar relative risk was also 1.25 [38]. The exact mechanisms through which passive smoking mediates all these detrimental effects to human vasculature have also been revealed during the last years. Nowadays it is well known that passive smoking leads to platelet activation, which is an important step in the cascade of formatting thrombus within the vessels [43]. In the same way, smoking enhances the secretion of the sympathetic system, which in turn acts on the endothelium, which finally reacts by secreting endothelin [44]. Endothelin is a well-known proinflammatory cytokine causing vasoconstriction on the small vessels [45]. Passive smoking causes direct endothelial damage [46–49], while it promotes the elimination of HDL, which is a compound that is beneficial for the vessels [50,51]. Moreover, passive smoking mediates the presence of insulin resistance, which precipitates to diabetes, a well-known major risk factor for cardiovascular disease, by enhancing the arterial stiffness, which in turn leads to arterial hypertension, another major risk factor for cardiovascular disease, that induces oxidative stress and increases the mean heart rate; all these actions play a key role in the initiation and progression of cardiovascular disease [52]. Various attempts at smoking cessation in public places have been implemented during recent years, showing a decrease in the incidence of fatal events; however, these efforts are only just beginning and newer studies must be done in order to specify and to organize these measures in a more effective way (Fig. 6.1).

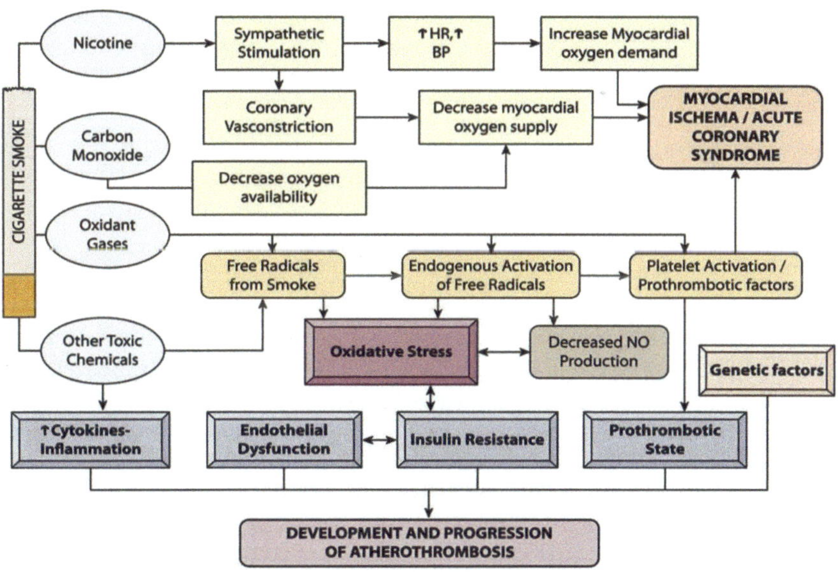

FIGURE 6.1 Pathophysiological effect of cigarette smoking and nicotine on cardiovascular disease. *HR*, Heart rate; *BP*, blood pressure; *NO*, nitrogen monoxide. (*Figure reprinted with permission from Salahuddin S, Prabhakaran D, Roy A. Pathophysiological mechanisms of tobacco-related CVD. Global Heart 2012;7(2):113–9.)*

METAL TOXICITY

Metal toxicity or metal poisoning is the toxic effect of certain metals in certain forms and doses on life. Some metals are toxic when they form poisonous soluble compounds. Certain metals have no biological role, ie, are not essential minerals, or are toxic when in a certain form. Potentially thousands of various metals and their compounds can cause damage to human homeostasis; in this section we refer to some of the heavy metals that mainly have been studied, including lead, cadmium, arsenic, and mercury.

Lead

Lead is the most common toxic element. Volcanic activity and geochemical weathering are the greatest natural sources. Lead-based paints, gasoline additives, food-can soldering, battery making, and soldered joints of drinking water pipe systems represent anthropogenic sources of lead in the environment [53]. An additional source is aircraft aviation engine fuel, while the tetraethyl lead as a gasoline additive in vehicles has been largely banned worldwide. The gastrointestinal absorption of lead is higher for children (30–50%) than for adults (5–10%). The absorbed lead is distributed to the blood, where red blood cells bind to almost 98% of lead. Gastrointestinal absorption and retention, the major pathway of lead intake, have been shown to vary widely depending on the chemical environment of the gastrointestinal lumen, person's age, and iron stores (nutritional status of the subject). Certain dietary components may act by increasing lead solubility, such as ascorbic acid, amino acids, vitamin D, protein, fat, and lactose, thus enhancing its absorption. Absorbed lead is mainly excreted in urine, whereas the feces contain predominantly unabsorbed lead. Being one of the calcium-like elements, lead follows the movement of calcium in the body to a large extent, and physiologic regulators of calcium metabolism usually affect the behavior of lead in a similar manner [54]. The exposure to lead is mainly defined from the lead levels in blood, however, this represents only a small part of the lead in the body, since larger amounts of this element are stored in soft tissues and bones. The ability to detect lead in urine is another means of defining the lead toxicity. A plethora of data suggest the deleterious effects of lead on the cardiovascular system. In the study of Weisskopf et al., it was shown that higher lead concentration in bones was strongly associated with all-cause mortality and mainly with cardiovascular mortality; while in another study, which evaluated the levels of lead in blood, those with a higher concentration of lead in blood were 1.4 times more vulnerable to cardiovascular mortality [55,56]. Menke et al. studied 13,946 adults from the NHAVES Group according to the lead values in blood and followed them up for 12 years. It was found that those in the upper tertile according to lead concentration in comparison with those in the lowest tertile were more prone to cardiovascular death, to myocardial infarction, and stroke [57]. There are many suggested mechanisms that mediate the actions of lead to vasculature, however, sufficient data underline the role of lead in hypertension and dyslipidemia, two classical risk factors for atherosclerosis [58–66].

Cadmium

Cadmium is a chemical element with symbol Cd and atomic number 48. This soft, bluish-white metal is chemically similar to the two other stable metals in group 12, zinc and mercury. It prefers oxidation, which means that it is mainly found as a +2 cation. Cadmium is considered as one of the most toxic metals because of its ubiquity and its long half-life [67]. It is found in food, vegetables, and water as well in tobacco smoke, since it is easily and preferably absorbed by the tobacco plants [67–70]. Cadmium enters the human body through the gastrointestinal tract, while evidence suggests that the absorption is greater among children. Iron is a key element in cadmium absorption since iron deficiency enhances cadmium absorbance, which in part explains the higher levels of cadmium among women [53]. Once absorbed, cadmium binds to metallothionein, which is a cysteine-rich protein that is mainly found in liver, kidney, and testes. This fact explains the high concentration of cadmium in these tissues, with a kidney predominance [53]. Cadmium can be measured in tissues as well as in blood, with the blood levels being the most representative for determining the cadmium toxicity [71]. Experimental data suggest a direct association of cadmium with atherosclerosis [72–74]. Similar conclusions are being made from evaluation of the exposure on humans. Menke et al. reported that an increase in cadmium urine level was accordingly associated not only with all-cause mortality but also with mortality from cardiovascular causes among men [70]. Other studies have reported similar findings [75,76]. In a meta-analysis based on data from 12 pooled studies, it was found that cadmium was associated with the incidence of cardiovascular disease in total as well as separately with coronary heart disease, peripheral arteriopathy, and stroke [77]. In the large epidemiological study from NHAVES Group, it was found that cadmium level is a risk factor for the incidence of arterial hypertension, an established risk factor for cardiovascular disease [78]. The mechanisms that mediate these phenomena are multiple. It seems that cadmium upregulates the release of cytokines and cell adhesion molecules, which in turn provoke endothelial activation and increase the endothelial cell permeability and death [79]. Another suggested mechanism is the induction of oxidative stress, which in turn provokes the necessary deleterious effects on the microvasculature [80–83].

Mercury

Mercury is widely known as quicksilver and was formerly named hydrargyrum. Common sources of mercury exposure include proximity to mercury mining sites, recycling facilities, medical or municipal incinerators, coal-fired power plants, or mercury-containing latex paint [84]. On average around the globe, there are indications that anthropogenic emissions of mercury have resulted in deposition rates today that are 1.5–3 times higher than those during preindustrial times. In and around industrial areas the deposition rates have increased by 2–10 times during the last 200 years. Mercury can also be found in seafood and in corn products; as well, evidence suggests that it is also present in various industrial products [84–86]. Dietary methylmercury enters the body via the gastrointestinal tract, immediately after absorption enters the blood, where it binds strongly to the red cells. Mercury is found in all tissues, about one-tenth is found in the brain and only 5% is found free in plasma. Mercury in the body binds to sulfur atoms almost exclusively and it is eliminated from feces; however, it has a long half-life of more than 70–80 days, which means practically that the concentration of this element in blood tends to be always increasing, since the human body has no ability to directly excrete mercury [87]. Data also support a direct association of mercury with cardiovascular disease. The connection between this element and the disease is based on in vivo as well as in vitro observations [88,89]. The first observation on this element was the Kuopio Ischemic Heart Disease Risk Factor study, where it was shown that the mercury intake and the consequent concentration in body was associated with higher prevalence of coronary heart disease, all-cause mortality, as well as the incidence of myocardial infarction [90]. As previously referred to, fish is the dominant source of mercury for the human body. Fish is an important source of necessary elements and its consumption is generally considered healthy. However, the deleterious effects of mercury can sometimes diminish the beneficial impact of the consumption of fish on human homeostasis [91]. Inconsistent observations exist on the professional exposure to mercury [92,93], while also studies referring to the role of mercury on hypertension have reached some controversial conclusions [94–96]. The mechanisms through which mercury mediates all these activities are not completely understood and explained, however, it seems that mercury induces oxidative stress, the depletion of sulfhydryl groups, as well as mitochondrial necrosis and consequently cellular apoptosis [97–99].

Arsenic

Arsenic occurs in nature as a monoisotopic element, composed of one stable isotope. It is a metalloid and can exist in various allotropes, with the gray form being the most important in industry. The earth's crust is an abundant natural source of arsenic. It is present in more than 200 different minerals, the most common of which is called arsenopyrite. About one-third of the arsenic in the earth's atmosphere is of natural origin. Volcanic action is the most important natural source. The next most important source is arsenic-containing vapor that is generated from solid or liquid forms of arsenic salts at low temperatures. Inorganic arsenic of geological origin is found in groundwater used as drinking water in several parts of the world, for example, Bangladesh, India, and Taiwan. Elemental arsenic is produced commercially from arsenic trioxide. Arsenic trioxide is a by-product of metal-smelting operations. About 70% of the world production of arsenic is used in timber treatment, 22% in agricultural chemicals, and the remainder in glass, pharmaceuticals, and metallic alloys. Mining, metal smelting, and burning of fossil fuels are the major industrial processes that contribute to arsenic contamination of air, water, and soil. The use of arsenic-containing pesticides in the past has left large areas of agricultural land contaminated [100]. Arsenic enters the human body via inhalation and the gastrointestinal tract. Once absorbed, arsenic is metabolized to form glutathione-arsenic complexes, which are simultaneously also subjected to methylation that occurs mainly in the liver producing water-soluble monomethylarsinic acid and dimethylarsinic acid, which in turn are eliminated through the urine. Since arsenic is quickly cleared from blood, urine is the best and most appropriate means of determining arsenic levels in humans [101]. The role of arsenic in cardiovascular disease is not as well studied as the prior discussed metals. Although epidemiological data suggest a clear negative association between arsenic levels in humans and the evolution of atherosclerosis, the data from studies based on occupational exposure are inconsistent [102]. In the large study from Moon et al., it was shown that the higher the concentration of arsenic in humans as measured from urine samples the greater the risk for cardiovascular as well as coronary artery disease mortality [103]. Other studies have shown that exposure through drinking water has led to an increasing incidence of adverse cardiovascular endpoints [104]. An ecological study, conducted in the arseniasis-endemic areas of southwestern Taiwan, reported increased age-adjusted mortality from ischemic heart diseases compared with residents in nonendemic areas [105]. Similarly, the exposure to arsenic was associated with the incidence of myocardial infarction in Chile [106]. The mechanisms mediating these actions are also complex. Firstly, the fact that arsenic can predispose to diabetes is an early explanation for these phenomena, since diabetes is a well-established risk factor for cardiovascular disease [107,108]. Another potential mechanism is the induction of proinflammatory cytokines, such as IL-6 and MCP-1, which in turn fuel the sequence of the events that lead to atherosclerosis [109,110]. Lastly but not of least importance, it is the capacity of arsenic to induce reactive oxygen species and thus the inflammatory cascade that promotes the initiation and evolution of atherosclerosis [110–114] (Table 6.2).

TABLE 6.2 Main Studies Associating Metal Toxicity and Cardiovascular Disease

Study Group	Metal	Study Design	Conclusions
Weisskopf et al. [55]	Lead	Association between bone (tibia and patella) lead levels and all cause as well as CV mortality in 868 men from normative Aging Study.	Compared to the lowest tertile of patella bone lead, the fully adjusted HR in the highest tertile for all-cause and CV mortality were 2.52 (95% CI: 1.17–5.41) and 5.63 (95% CI: 1.73–18.3), respectively. Results were similar for tibia lead.
Menke et al. [57]	Lead	Association of blood lead levels with all-cause mortality and CV death in 13946 adults from the Third National Health and Nutrition Examination Survey recruited in 1988–1994 and followed up for up to 12 years.	After multivariate adjustment, the hazard ratios (95% CI) for comparisons of participants in the highest tertile of blood lead with those in the lowest tertile were 1.25 (1.04–1.51, $p=0.002$) for all-cause mortality and 1.55 (1.08 to 2.55, $p=0.003$) for CV mortality. Blood lead level was significantly associated with both MI and stroke mortality.
Menke et al. [70]	Cadmium	Relationship between cadmium measured in 13,958 adults from the Third National Health and Nutrition Examination Survey, who were followed for 6 years, and all-cause, CVD, and CHD mortality.	After multivariable adjustment, including smoking, a major source of cadmium exposure in nonoccupationally exposed populations, the hazard ratios [95% confidence interval (CI)] for all-cause, CVD, and CHD mortality associated with a twofold higher creatinine-corrected urinary cadmium were, respectively: 1.28 (95% CI, 1.15–1.43), 1.21 (95% CI, 1.07–1.36), and 1.36 (95% CI, 1.11–1.66) for men; and 1.06 (95% CI, 0.96–1.16), 1.07 (95% CI, 0.85–1.35), 0.93 (95% CI, 0.84–1.04), and 0.82 (95% CI, 0.76–0.89) for women.

Tellez-Plaza et al. [77]	Cadmium	Meta-analysis of 12 pooled studies. Study of the association between cadmium levels and all-cause mortality, CV mortality, CHD, PAD, and stroke.	The pooled relative risks (95% confidence interval) for CVD, CHD, stroke, and PAD were: 1.36 (95%CI: 1.11, 1.66), 1.30 (95%CI: 1.12, 1.52), 1.18 (95%CI: 0.86, 1.59), and 1.49 (95%CI: 1.15, 1.92), respectively. The pooled relative risks for CVD in men, women, and never smokers were 1.29 (1.12, 1.48), 1.20 (0.92, 1.56), and 1.27 (0.97, 1.67), respectively.
Virtanen et al. [90]	Mercury	1871 Finnish men aged 42 to 60 without previous CVD. Follow-up for 14 years. Endpoints: all-cause mortality, acute coronary events, CHD, CVD.	Men in the highest third of hair mercury content had an adjusted 1.60-fold (95% CI, 1.24–2.06) risk of acute coronary event, 1.68-fold (95% CI, 1.15–2.44) risk of CVD, 1.56-fold (95% CI, 0.99–2.46) risk of CHD, and 1.38-fold (95% CI, 1.15–1.66) risk of any death compared with men in the lower two-thirds.
Moon et al. [103]	Arsenic	3575 American Indian men and women aged 45–74 years living in Arizona, Oklahoma, and North and South Dakota, followed up for 20 years. Endpoints: CVD, CHD, stroke mortality.	Comparing the highest to lowest quartile arsenic concentration the hazard ratios (95% confidence interval) for CVD, CHD, and stroke mortality after adjustment for all related risk factors were 1.65 (1.20, 2.27; p-trend $<$ 0.001), 1.71 (1.19, 2.44; p-trend $<$ 0.001), and 3.03 (1.08, 8.50; p-trend = 0.061), respectively. The corresponding hazard ratios for incident CVD, CHD, and stroke were 1.32 (1.09, 1.59; p-trend = 0.002), 1.30 (1.04, 1.62; p-trend = 0.006), and 1.47 (0.97, 2.21; p-trend = 0.032), respectively.

CVD, Cardiovascular disease; *CHD*, coronary heart disease; *PAD*, peripheral arterial disease; *CV*, cardiovascular; *MI*, myocardial infarction.

CONCLUSIONS

Cardiovascular disease remains the first cause of mortality in all of the developed world. Cardiovascular disease is a complex mixture of diverse mechanisms, in which genetic and environmental factors integrate in order to produce a result that is detrimental for human health. While the genetic part cannot as of yet be influenced, the environmental factors are important to consider in dealing with cardiovascular disease. A plethora of factors mediate their actions and result generally termed environmental influences on cardiovascular health. These factors are so variable, and they also interplay with each other, which practically means that a detailed presentation in a restricted framework would be impossible. Many of these environmental factors are not known, underlying mechanisms are not well understood, and newer studies have to be undertaken in order to better understand this field. This is possibly the most important step in order to face this scourge of humanity—cardiovascular disease.

REFERENCES

[1] Marks GB. A critical appraisal of the evidence for adverse respiratory effects due to exposure to environmental ozone and particulate pollution: relevance to air quality guidelines. Aust N Z J Med 1994;24(2):202–13.

[2] Lee BJ, Kim B, Lee K. Air pollution exposure and cardiovascular disease. Toxicol Res 2014;30(2):71–5.

[3] Gold DR, Litonjua A, Schwartz J, Lovett E, Larson A, Nearing B, et al. Ambient pollution and heart rate variability. Circulation 2000;101(11):1267–73.

[4] Pope 3rd CA, Burnett RT, Thurston GD, Thun MJ, Calle EE, Krewski D, et al. Cardiovascular mortality and long-term exposure to particulate air pollution: epidemiological evidence of general pathophysiological pathways of disease. Circulation 2004;109(1):71–7.

[5] Wellenius GA, Schwartz J, Mittleman MA. Particulate air pollution and hospital admissions for congestive heart failure in seven United States cities. Am J Cardiol 2006;97(3): 404–8.

[6] Pope 3rd CA, Muhlestein JB, May HT, Renlund DG, Anderson JL, Horne BD. Ischemic heart disease events triggered by short-term exposure to fine particulate air pollution. Circulation 2006;114(23):2443–8.

[7] Samet JM, Dominici F, Curriero FC, Coursac I, Zeger SL. Fine particulate air pollution and mortality in 20 U.S. cities, 1987–1994. N Engl J Med 2000;343(24):1742–9.

[8] Brook RD. Cardiovascular effects of air pollution. Clin Sci (Lond) 2008;115(6):175–87.

[9] Brook RD, Brook JR, Rajagopalan S. Air pollution: the "Heart" of the problem. Curr Hypertens Rep 2003;5(1):32–9.

[10] Zanobetti A, Schwartz J, Samoli E, Gryparis A, Touloumi G, Peacock J, et al. The temporal pattern of respiratory and heart disease mortality in response to air pollution. Environ Health Perspect 2003;111(9):1188–93.

[11] Morris RD. Airborne particulates and hospital admissions for cardiovascular disease: a quantitative review of the evidence. Environ Health Perspect 2001;109(Suppl.):4495–500.

[12] Zanobetti A, Schwartz J. The effect of particulate air pollution on emergency admissions for myocardial infarction: a multicity case-crossover analysis. Environ Health Perspect 2005;113(8):978–82.

[13] Low RB, Bielory L, Qureshi AI, Dunn V, Stuhlmiller DF, Dickey DA. The relation of stroke admissions to recent weather, airborne allergens, air pollution, seasons, upper respiratory infections, and asthma incidence, September 11, 2001, and day of the week. Stroke 2006;37(4):951–7.

[14] Dominici F, Peng RD, Bell ML, Pham L, McDermott A, Zeger SL, et al. Fine particulate air pollution and hospital admission for cardiovascular and respiratory diseases. JAMA 2006;295(10):1127–34.

[15] Grahame TJ, Schlesinger RB. Oxidative stress-induced telomeric erosion as a mechanism underlying airborne particulate matter-related cardiovascular disease. Part Fibre Toxicol 2012;9:21.

[16] Brook RD, Brook JR, Urch B, Vincent R, Rajagopalan S, Silverman F. Inhalation of fine particulate air pollution and ozone causes acute arterial vasoconstriction in healthy adults. Circulation 2002;105(13):1534–6.

[17] Zanobetti A, Canner MJ, Stone PH, Schwartz J, Sher D, Eagan-Bengston E, et al. Ambient pollution and blood pressure in cardiac rehabilitation patients. Circulation 2004;110(15):2184–9.

[18] Tsuji H, Venditti Jr FJ, Manders ES, Evans JC, Larson MG, Feldman CL, et al. Reduced heart rate variability and mortality risk in an elderly cohort. The Framingham Heart Study. Circulation 1994;90(2):878–83.

[19] Tsuji H, Larson MG, Venditti Jr FJ, Manders ES, Evans JC, Feldman CL, et al. Impact of reduced heart rate variability on risk for cardiac events. The Framingham Heart Study. Circulation 1996;94(11):2850–5.

[20] Touloumi G, Katsouyanni K, Zmirou D, Schwartz J, Spix C, de Leon AP, et al. Short-term effects of ambient oxidant exposure on mortality: a combined analysis within the APHEA project. Air pollution and health: a European approach. Am J Epidemiol 1997;146(2):177–85.

[21] Odemuyiwa O, Malik M, Farrell T, Bashir Y, Poloniecki J, Camm J. Comparison of the predictive characteristics of heart rate variability index and left ventricular ejection fraction for all-cause mortality, arrhythmic events and sudden death after acute myocardial infarction. Am J Cardiol 1991;68(5):434–9.

[22] Kodavanti UP, Schladweiler MC, Ledbetter AD, Watkinson WP, Campen MJ, Winsett DW, et al. The spontaneously hypertensive rat as a model of human cardiovascular disease: evidence of exacerbated cardiopulmonary injury and oxidative stress from inhaled emission particulate matter. Toxicol Appl Pharmacol 2000;164(3):250–63.

[23] Bocci V, Valacchi G, Corradeschi F, Aldinucci C, Silvestri S, Paccagnini E, et al. Studies on the biological effects of ozone: 7. Generation of reactive oxygen species (ROS) after exposure of human blood to ozone. J Biol Regul Homeost Agents 1998;12(3):67–75.

[24] Bocci V, Paulesu L. Studies on the biological effects of ozone 1. Induction of interferon gamma on human leukocytes. Haematologica 1990;75(6):510–5.

[25] Bocci V, Luzzi E, Corradeschi F, Paulesu L, Rossi R, Cardaioli E, et al. Studies on the biological effects of ozone: 4. Cytokine production and glutathione levels in human erythrocytes. J Biol Regul Homeost Agents 1993;7(4):133–8.

[26] Le Tertre A, Quenel P, Eilstein D, Medina S, Prouvost H, Pascal L, et al. Short-term effects of air pollution on mortality in nine French cities: a quantitative summary. Arch Environ Health 2002;57(4):311–9.

[27] Gryparis A, Forsberg B, Katsouyanni K, Analitis A, Touloumi G, Schwartz J, et al. Acute effects of ozone on mortality from the "air pollution and health: a European approach" project. Am J Respir Crit Care Med 2004;170(10):1080–7.

[28] Goldberg MS, Burnett RT, Brook J, Bailar 3rd JC, Valois MF, Vincent R. Associations between daily cause-specific mortality and concentrations of ground-level ozone in Montreal, Quebec. Am J Epidemiol 2001;154(9):817–26.
[29] Ruidavets JB, Cournot M, Cassadou S, Giroux M, Meybeck M, Ferrieres J. Ozone air pollution is associated with acute myocardial infarction. Circulation 2005;111(5):563–9.
[30] Koken PJ, Piver WT, Ye F, Elixhauser A, Olsen LM, Portier CJ. Temperature, air pollution, and hospitalization for cardiovascular diseases among elderly people in Denver. Environ Health Perspect 2003;111(10):1312–7.
[31] Sunyer J, Castellsague J, Saez M, Tobias A, Anto JM. Air pollution and mortality in Barcelona. J Epidemiol Community Health 1996;50(Suppl.):76–80.
[32] Mons U, Muezzinler A, Gellert C, Schottker B, Abnet CC, Bobak M, et al. Impact of smoking and smoking cessation on cardiovascular events and mortality among older adults: meta-analysis of individual participant data from prospective cohort studies of the CHANCES consortium. BMJ 2015;350:h1551.
[33] Pujades-Rodriguez M, George J, Shah AD, Rapsomaniki E, Denaxas S, West R, et al. Heterogeneous associations between smoking and a wide range of initial presentations of cardiovascular disease in 1937360 people in England: lifetime risks and implications for risk prediction. Int J Epidemiol 2015;44(1):129–41.
[34] Glantz SA, Parmley WW. Passive smoking and heart disease. Epidemiology, physiology, and biochemistry. Circulation 1991;83(1):1–12.
[35] Glantz SA, Parmley WW. Passive smoking and heart disease. Mechanisms and risk. JAMA 1995;273(13):1047–53.
[36] Taylor AE, Johnson DC, Kazemi H. Environmental tobacco smoke and cardiovascular disease. A position paper from the council on cardiopulmonary and critical care, American Heart Association. Circulation 1992;86(2):699–702.
[37] Law MR, Morris JK, Wald NJ. Environmental tobacco smoke exposure and ischaemic heart disease: an evaluation of the evidence. BMJ 1997;315(7114):973–80.
[38] He J, Vupputuri S, Allen K, Prerost MR, Hughes J, Whelton PK. Passive smoking and the risk of coronary heart disease–a meta-analysis of epidemiologic studies. N Engl J Med 1999;340(12):920–6.
[39] Oono IP, Mackay DF, Pell JP. Meta-analysis of the association between secondhand smoke exposure and stroke. J Public Health (Oxf) 2011;33(4):496–502.
[40] Sebrie EM, Sandoya E, Bianco E, Hyland A, Cummings KM, Glantz SA. Hospital admissions for acute myocardial infarction before and after implementation of a comprehensive smoke-free policy in Uruguay: experience through 2010. Tob Control 2014;23(6):471–2.
[41] Fernandez de Bobadilla J, Dalmau R, Galve E, Legislation on Tobacco and Acute Coronary Syndrome in Spain Group. Laws restricting smoking in public places reduce the incidence of acute coronary syndrome in Spain. Rev Esp Cardiol (Engl Ed) 2014;67(5):349–52.
[42] Jan C, Lee M, Roa R, Herrera V, Politis M, Motta J. The association of tobacco control policies and the risk of acute myocardial infarction using hospital admissions data. PLoS One 2014;9(2):e88784.
[43] Davis JW, Shelton L, Watanabe IS, Arnold J. Passive smoking affects endothelium and platelets. Arch Intern Med 1989;149(2):386–9.
[44] Puranik R, Celermajer DS. Smoking and endothelial function. Prog Cardiovasc Dis 2003;45(6):443–58.
[45] Agapitov AV, Haynes WG. Role of endothelin in cardiovascular disease. J Renin Angiotensin Aldosterone Syst 2002;3(1):1–15.

[46] Lee TY, Gotlieb AI. Microfilaments and microtubules maintain endothelial integrity. Microsc Res Tech 2003;60(1):115–27.
[47] Cucina A, Sapienza P, Borrelli V, Corvino V, Foresi G, Randone B, et al. Nicotine reorganizes cytoskeleton of vascular endothelial cell through platelet-derived growth factor BB. J Surg Res 2000;92(2):233–8.
[48] Cucina A, Sapienza P, Corvino V, Borrelli V, Mariani V, Randone B, et al. Nicotine-induced smooth muscle cell proliferation is mediated through bFGF and TGF-beta 1. Surgery 2000;127(3):316–22.
[49] Cucina A, Sapienza P, Corvino V, Borrelli V, Randone B, Santoro-D'Angelo L, et al. Nicotine induces platelet-derived growth factor release and cytoskeletal alteration in aortic smooth muscle cells. Surgery 2000;127(1):72–8.
[50] Azizi F, Raiszadeh F, Salehi P, Rahmani M, Emami H, Ghanbarian A, et al. Determinants of serum HDL-C level in a Tehran urban population: the Tehran lipid and glucose study. Nutr Metab Cardiovasc Dis 2002;12(2):80–9.
[51] Mizoue T, Ueda R, Hino Y, Yoshimura T. Workplace exposure to environmental tobacco smoke and high density lipoprotein cholesterol among nonsmokers. Am J Epidemiol 1999;150(10):1068–72.
[52] Barnoya J, Glantz SA. Cardiovascular effects of secondhand smoke: nearly as large as smoking. Circulation 2005;111(20):2684–98.
[53] Solenkova NV, Newman JD, Berger JS, Thurston G, Hochman JS, Lamas GA. Metal pollutants and cardiovascular disease: mechanisms and consequences of exposure. Am Heart J 2014;168(6):812–22.
[54] Alissa EM, Ferns GA. Heavy metal poisoning and cardiovascular disease. J Toxicol 2011; 2011: 21. Article ID: 870125.
[55] Weisskopf MG, Jain N, Nie H, Sparrow D, Vokonas P, Schwartz J, et al. A prospective study of bone lead concentration and death from all causes, cardiovascular diseases, and cancer in the Department of Veterans Affairs Normative Aging Study. Circulation 2009;120(12): 1056–64.
[56] Lustberg M, Silbergeld E. Blood lead levels and mortality. Arch Intern Med 2002; 162(21):2443–9.
[57] Menke A, Muntner P, Batuman V, Silbergeld EK, Guallar E. Blood lead below 0.48 micromol/L (10 microg/dL) and mortality among US adults. Circulation 2006;114(13): 1388–94.
[58] Wannamethee SG, Shaper AG, Walker M. Changes in physical activity, mortality, and incidence of coronary heart disease in older men. Lancet 1998;351(9116):1603–8.
[59] Appel LJ, Moore TJ, Obarzanek E, Vollmer WM, Svetkey LP, Sacks FM, et al. A clinical trial of the effects of dietary patterns on blood pressure. DASH Collaborative Research Group. N Engl J Med 1997;336(16):1117–24.
[60] Clarke R, Frost C, Collins R, Appleby P, Peto R. Dietary lipids and blood cholesterol: quantitative meta-analysis of metabolic ward studies. BMJ 1997;314(7074):112–7.
[61] Danesh J, Collins R, Peto R. Lipoprotein(a) and coronary heart disease. Meta-analysis of prospective studies. Circulation 2000;102(10):1082–5.
[62] Blake GJ, Ridker PM. Novel clinical markers of vascular wall inflammation. Circ Res 2001;89(9):763–71.
[63] Beattie JH, Kwun IS. Is zinc deficiency a risk factor for atherosclerosis? Br J Nutr 2004;91(2):177–81.
[64] Brenneisen P, Steinbrenner H, Sies H. Selenium, oxidative stress, and health aspects. Mol Aspects Med 2005;26(4–5):256–67.

[65] Hummel M, Standl E, Schnell O. Chromium in metabolic and cardiovascular disease. Horm Metab Res 2007;39(10):743–51.
[66] Gallagher CM, Meliker JR. Blood and urine cadmium, blood pressure, and hypertension: a systematic review and meta-analysis. Environ Health Perspect 2010;118(12):1676–84.
[67] Nawrot TS, Staessen JA, Roels HA, Munters E, Cuypers A, Richart T, et al. Cadmium exposure in the population: from health risks to strategies of prevention. Biometals 2010;23(5):769–82.
[68] Wagner GJ, Yeargan R. Variation in cadmium accumulation potential and tissue distribution of cadmium in tobacco. Plant Physiol 1986;82(1):274–9.
[69] Jarup L, Berglund M, Elinder CG, Nordberg G, Vahter M. Health effects of cadmium exposure–a review of the literature and a risk estimate. Scand J Work Environ Health 1998;24(Suppl):11–51.
[70] Menke A, Muntner P, Silbergeld EK, Platz EA, Guallar E. Cadmium levels in urine and mortality among U.S. adults. Environ Health Perspect 2009;117(2):190–6.
[71] Nordberg GF, Nogawa K, Nordberg M, et al. Cadmium. In: Nordberg GF, Fowler BA, Nordberg M, Friberg LT, editors. Handbook on toxicology of metals. 3rd ed. Elsevier; 2007.
[72] Knoflach M, Messner B, Shen YH, Frotschnig S, Liu G, Pfaller K, et al. Non-toxic cadmium concentrations induce vascular inflammation and promote atherosclerosis. Circ J 2011;75(10): 2491–5.
[73] Ronco AM, Montenegro M, Castillo P, Urrutia M, Saez D, Hirsch S, et al. Maternal exposure to cadmium during gestation perturbs the vascular system of the adult rat offspring. Toxicol Appl Pharmacol 2011;251(2):137–45.
[74] Messner B, Knoflach M, Seubert A, Ritsch A, Pfaller K, Henderson B, et al. Cadmium is a novel and independent risk factor for early atherosclerosis mechanisms and in vivo relevance. Arterioscler Thromb Vasc Biol 2009;29(9):1392–8.
[75] Tellez-Plaza M, Navas-Acien A, Menke A, Crainiceanu CM, Pastor-Barriuso R, Guallar E. Cadmium exposure and all-cause and cardiovascular mortality in the U.S. general population. Environ Health Perspect 2012;120(7):1017–22.
[76] Tellez-Plaza M, Guallar E, Howard BV, Umans JG, Francesconi KA, Goessler W, et al. Cadmium exposure and incident cardiovascular disease. Epidemiology 2013;24(3):421–9.
[77] Tellez-Plaza M, Jones MR, Dominguez-Lucas A, Guallar E, Navas-Acien A. Cadmium exposure and clinical cardiovascular disease: a systematic review. Curr Atheroscler Rep 2013;15(10):356.
[78] Tellez-Plaza M, Navas-Acien A, Crainiceanu CM, Guallar E. Cadmium exposure and hypertension in the 1999–2004 National health and nutrition examination survey (NHANES). Environ Health Perspect 2008;116(1):51–6.
[79] Messner B, Bernhard D. Cadmium and cardiovascular diseases: cell biology, pathophysiology, and epidemiological relevance. Biometals 2010;23(5):811–22.
[80] Valko M, Morris H, Cronin MT. Metals, toxicity and oxidative stress. Curr Med Chem 2005;12(10):1161–208.
[81] Prozialeck WC, Edwards JR, Nebert DW, Woods JM, Barchowsky A, Atchison WD. The vascular system as a target of metal toxicity. Toxicol Sci 2008;102(2):207–18.
[82] Varoni MV, Palomba D, Satta M, Satta A, Anania V. Different response of intracerebroventricular cadmium administration on blood pressure in normal and low urinary kallikrein rats. Pharmacol Res 2003;47(2):157–62.
[83] Satarug S, Nishijo M, Lasker JM, Edwards RJ, Moore MR. Kidney dysfunction and hypertension: role for cadmium, p450 and heme oxygenases? Tohoku J Exp Med 2006;208(3): 179–202.

[84] Berlin M, Zalups RK, Fowler BA. Mercury. In: Nordberg GF, Fowler BA, Nordberg M, Friberg LT, editors. Handbook on the toxicology of metals. 3rd ed. Elsevier; 2007. p. 675–729.
[85] Mercury in canned tuna still a concern: new tests reinforce a need for some people to limit consumption. Consum Rep 2011:20–1.
[86] Agency for Toxic Substances and Disease Registry (ATSDR). Toxicological profiles. Toxicological profile for mercury. Atlanta (GA): U.S. Department of Health and Human Services, Public Health Service; 1999.
[87] Diez S. Human health effects of methylmercury exposure. Rev Environ Contam Toxicol 2009;1981:132–7.
[88] Lund BO, Miller DM, Woods JS. Studies on Hg(II)-induced H_2O_2 formation and oxidative stress in vivo and in vitro in rat kidney mitochondria. Biochem Pharmacol 1993;45(10):2017–24.
[89] Huang YL, Cheng SL, Lin TH. Lipid peroxidation in rats administrated with mercuric chloride. Biol Trace Elem Res 1996;52(2):193–206.
[90] Virtanen JK, Voutilainen S, Rissanen TH, Mursu J, Tuomainen TP, Korhonen MJ, et al. Mercury, fish oils, and risk of acute coronary events and cardiovascular disease, coronary heart disease, and all-cause mortality in men in eastern Finland. Arterioscler Thromb Vasc Biol 2005;25(1):228–33.
[91] Guallar E, Sanz-Gallardo MI, van't Veer P, Bode P, Aro A, Gomez-Aracena J, et al. Mercury, fish oils, and the risk of myocardial infarction. N Engl J Med 2002;347(22):1747–54.
[92] Yoshizawa K, Rimm EB, Morris JS, Spate VL, Hsieh CC, Spiegelman D, et al. Mercury and the risk of coronary heart disease in men. N Engl J Med 2002;347(22):1755–60.
[93] Hallgren CG, Hallmans G, Jansson JH, Marklund SL, Huhtasaari F, Schutz A, et al. Markers of high fish intake are associated with decreased risk of a first myocardial infarction. Br J Nutr 2001;86(3):397–404.
[94] Salonen JT, Seppanen K, Lakka TA, Salonen R, Kaplan GA. Mercury accumulation and accelerated progression of carotid atherosclerosis: a population-based prospective 4-year follow-up study in men in Eastern Finland. Atherosclerosis 2000;148(2):265–73.
[95] Valera B, Dewailly E, Poirier P. Environmental mercury exposure and blood pressure among Nunavik Inuit adults. Hypertension 2009;54(5):981–6.
[96] Vupputuri S, Batuman V, Muntner P, Bazzano LA, Lefante JJ, Whelton PK, et al. Effect of blood pressure on early decline in kidney function among hypertensive men. Hypertension 2003;42(6):1144–9.
[97] Yonaha M, Saito M, Sagai M. Stimulation of lipid peroxidation by methyl mercury in rats. Life Sci 1983;32(13):1507–14.
[98] Clarkson TW. The pharmacology of mercury compounds. Annu Rev Pharmacol 1972:12375–406.
[99] InSug O, Datar S, Koch CJ, Shapiro IM, Shenker BJ. Mercuric compounds inhibit human monocyte function by inducing apoptosis: evidence for formation of reactive oxygen species, development of mitochondrial membrane permeability transition and loss of reductive reserve. Toxicology 1997;124(3):211–24.
[100] www.greenfacts.org.
[101] Fowler BA, Selene CH,Chou J, et al. Arsenic. In: Handbook on toxicology of metals. 3rd ed. Elsevier; 2007. p. 367–406.
[102] National Research Council. Arsenic in drinking water. Washington, DC, USA: National Academy Press; 1999.
[103] Moon KA, Guallar E, Umans JG, Devereux RB, Best LG, Francesconi KA, et al. Association between exposure to low to moderate arsenic levels and incident cardiovascular disease. A prospective cohort study. Ann Intern Med 2013;159(10):649–59.

[104] Zierold KM, Knobeloch L, Anderson H. Prevalence of chronic diseases in adults exposed to arsenic-contaminated drinking water. Am J Public Health 2004;94(11):1936–7.

[105] Tsai SM, Wang TN, Ko YC. Mortality for certain diseases in areas with high levels of arsenic in drinking water. Arch Environ Health 1999;54(3):186–93.

[106] Yuan Y, Marshall G, Ferreccio C, Steinmaus C, Selvin S, Liaw J, et al. Acute myocardial infarction mortality in comparison with lung and bladder cancer mortality in arsenic-exposed region II of Chile from 1950 to 2000. Am J Epidemiol 2007;166(12):1381–91.

[107] Patlolla AK, Tchounwou PB. Serum acetyl cholinesterase as a biomarker of arsenic induced neurotoxicity in sprague-dawley rats. Int J Environ Res Public Health 2005;2(1):80–3.

[108] Tseng CH, Tai TY, Chong CK, Tseng CP, Lai MS, Lin BJ, et al. Long-term arsenic exposure and incidence of non-insulin-dependent diabetes mellitus: a cohort study in arseniasis-hyperendemic villages in Taiwan. Environ Health Perspect 2000;108(9):847–51.

[109] Lee PC, Ho IC, Lee TC. Oxidative stress mediates sodium arsenite-induced expression of heme oxygenase-1, monocyte chemoattractant protein-1, and interleukin-6 in vascular smooth muscle cells. Toxicol Sci 2005;85(1):541–50.

[110] Wu MM, Chiou HY, Ho IC, Chen CJ, Lee TC. Gene expression of inflammatory molecules in circulating lymphocytes from arsenic-exposed human subjects. Environ Health Perspect 2003;111(11):1429–38.

[111] Lee TC, Ho IC. Modulation of cellular antioxidant defense activities by sodium arsenite in human fibroblasts. Arch Toxicol 1995;69(7):498–504.

[112] Wu MM, Chiou HY, Wang TW, Hsueh YM, Wang IH, Chen CJ, et al. Association of blood arsenic levels with increased reactive oxidants and decreased antioxidant capacity in a human population of northeastern Taiwan. Environ Health Perspect 2001;109(10):1011–7.

[113] Chen Y, Santella RM, Kibriya MG, Wang Q, Kappil M, Verret WJ, et al. Association between arsenic exposure from drinking water and plasma levels of soluble cell adhesion molecules. Environ Health Perspect 2007;115(10):1415–20.

[114] Ridker PM, Hennekens CH, Roitman-Johnson B, Stampfer MJ, Allen J. Plasma concentration of soluble intercellular adhesion molecule 1 and risks of future myocardial infarction in apparently healthy men. Lancet 1998;351(9096):88–92.

Chapter 7

Genetic Susceptibility to Cardiovascular Diseases: From Mendelian Disorders to Common Variants

A. Briasoulis
Wayne State University/Detroit Medical Center, Detroit, MI, United States

INTRODUCTION

Cardiovascular disease (CVD) is a leading health problem, affecting over 80,000,000 individuals in the United States alone. CVD encompasses a broad range of disorders including diseases of the vasculature, the myocardium, the heart's electrical system, and congenital heart disease. For nearly all of these disorders, inherited DNA sequence variants play a role in conferring risk for disease. For example, in the general population, a history of premature atherosclerotic CVD in a parent confers about a threefold increase in CVD risk to offspring [1]. The precise magnitude of the role of inheritance, however, varies by disease and by other factors such as age of disease onset and subtype of disease. Although it is usually stated that 20–50% of patients with coronary artery disease (CAD) lack conventional risk factors, large studies have reported that 85–90% of patients with CAD have one or more of the traditional risk factors [1,2]. The cross-sectional case–control INTERHEART study reported that 90% of CAD events could be explained on the basis of nine modifiable risk factors [3]. Nevertheless, numerous early case–control studies suggested approximately two- to fivefold higher prevalence of a positive familial history among subjects with manifest CAD than among control subjects [4–6]. Large, prospective cohort studies also have found a positive association between self-reported parental or familial history and multivariable-adjusted relative risks for offspring CVD, with estimates ranging from 0.8 to 2.2 [7]. Later prospective studies using validated instead of self-reported events demonstrated that familiar and sibling CAD history is a robust predictor of CAD, even after adjustment for all known risk factors [8,9], which suggests that there is a strong genetic

Cardiovascular Diseases.

component underlying this complex disease [8]. Genetic contribution toward CAD development has previously been validated in twin studies [10–12].

GENETIC STUDIES: A METHODOLOGIC APPROACH

The concept of genetic determinism in the case of CVD is mitigated by the impact of gene–environmental interactions. The phenotypic expression of complex genetic traits relies on age, gender, food and drug consumption, physical activity, and a variety of other factors. In such complex traits, once the genetic contribution has been evident, two main approaches have been applied to decipher the genetic background: (Genome-wide and fine-mapping) linkage analysis and (candidate gene or genome-wide) association analysis studies. The choice of approach has depended on the pattern of segregation, whether consistent with the ratios described by Mendel or more complex. Some forms of CVD exhibit a simple pattern of inheritance suggestive of a single causal gene that confers a large effect on phenotype. For many of these Mendelian forms of CVD, direct DNA sequencing and/or linkage analysis has successfully yielded the causal gene and mutation. Linkage analysis was used to localize the chromosomal position of a causal gene for homozygous familial hypercholesterolemia, the beta cardiac myosin heavy chain for hypertrophic cardiomyopathy, long QT syndrome, severe hypercholesterolemia, Mendelian forms of hypertension, Marfan syndrome, and several forms of congenital heart disease including septal defects and valve defects. However, most CVD traits such as myocardial infarction (MI) or concentrations of plasma low-density lipoprotein (LDL) cholesterol show complex inheritance, suggestive of an interplay between multiple genes and nongenetic factors. Mapping gene loci associated with complex traits requires substantial levels of information and analysis, but since 2007 approaches to accomplish this goal have matured, and genetic mapping for complex traits in humans has become a reality.

The key concepts behind linkage analysis are linkage and linkage disequilibrium (see Glossary). Once polymorphic markers across the genome of family members have been identified, the estimated logarithm of the odds (LOD) score represents the probability that a marker and putative causal locus are linked in pedigrees as opposed to the null hypothesis of the absence of linkage in the pedigrees. Logarithm of the odds scores >3 indicates significant linkage ($P < .0001$). The initial linkage is validated by fine mapping linkage analyses (addition of markers to form a dense genotype set in the region of linkage). The limited success of linkage analysis to identify a reproducible genetic locus so far is at least in part due to studies being too small to detect genes of modest effect.

Association studies are designed to compare allele or haplotype frequencies between case and control groups; a statistical difference in frequencies between cases and controls offers evidence that the allele or haplotype (combination of polymorphic alleles that are located at closely linked loci and tend to be inherited together) is associated with the trait. Candidate genes for association studies are selected on the basis of linkage study outcomes and prior biological knowledge of gene function. In the case of CAD association studies are based the common disease/common variant (CDCV) hypothesis [13,14]. Under this model, susceptibility to CAD results from the interaction of multiple common variants (minor allele frequency; MAF > 1%), most of which are shared by unrelated affected individuals and usually lack functional association with the disease, but may unveil functional variant that is in linkage disequilibrium (LD) with the observed variant. So far, whole genome association studies have been using single nucleotide polymorphisms (SNPs) with MAF > 5%. In addition, common variants exhibit odds ratio (OR) up to 1.5–2 [15]. On the other hand, the multiple rare-variant hypothesis, also referred to as the genetic heterogeneity model [16], suggests that disease susceptibility is because of distinct genetic variants with low population frequencies (frequencies <1%) in different individuals. Contrary to common variants, rare variants exhibit greater OR (>2%), functional relevance with the disease, and higher penetrance [15]. Genetic heterogeneity renders an association study expensive due to the large sample needed. Independently from the selected model approach, lack of reproducibility appears to be an undisputed limitation of association studies. This lack of reproducibility is often ascribed to inadequate statistical power, genetic and phenotypic heterogeneity (various clinical presentations of CAD), population-specific linkage disequilibrium, population stratification, and SNPs genotyped [17,18].

GENETIC POLYMORPHISMS IN CARDIOVASCULAR DISEASES

In most cases, genetic associations found in one study are not supported by subsequent studies. The lack of consistency in findings may be due to chance, underestimation of the frequency of polymorphisms in the control populations, not matching the race/ethnicity of cases and controls, or publication bias. Recently, Morgan et al. were unable to confirm as risk factors for acute coronary syndromes (ACSs) 85 genetic variants because none of them were validated in a large case–control study of 1461 participants [19]. Schematically, the spectrum of candidate genes in CAD comprises four components: endothelial injury, inflammation, coagulation/fibrinolysis, and lipid metabolism. The following section discusses the findings of the most important candidate gene studies of CAD.

Polymorphisms of Lipid Metabolism (Table 7.1)

Apolipoprotein E (apoE) is an LDL receptor ligand encoded by the APOE gene. In populations of European ancestry, three alleles ε2, ε3, and ε4, which have frequencies of 0.1, 0.75, and 0.13, respectively, correspond to three apoE variants. These result in significant differences in total cholesterol and LDL levels with ε2 carriers having the lowest and ε4 carriers the highest levels. The

TABLE 7.1 Polymorphisms Associated With Lipid Metabolism, Endothelial Function, Thrombosis/Fibrinolysis, and Inflammation

Study	Gene/ Polymorphism	Protein Encoded	Risk Genotype	Patients (Number)	Associated With	OR (CI) for ACS
Schmitz et al. [20]	ApoE/ε2,ε3,ε4 alleles	ApoE	ε4 allele	MI (1817)	Lipid metabolism	1.75 (1.19–2.57)
Tatsuguchi et al. [22]	OLR1/501G>C	OLR1	GC, CC	MI/controls (102/102)	Lipid metabolism	2.89 (1.51–5.53)
Liu et al. [24]	CETP/TaqIB	CETP	No difference	MI/controls (384/384)	Lipid metabolism	RR=0.95 (0.54–1.66)
Wittrup et al. [23]	LPL/Gly188Glu, Asp9Asn, Asn291Ser, and Ser447Ter	LPL	Gly 188Glu Ser447Ter (protective)	Ischemic heart disease (meta-analysis of 29 studies– 20,903 patients)	Lipid metabolism	4.9 (1.2–20) 0.8 (0.7–1.0)
Fatini et al. [29]	eNOS/894G>T, 786T>C, 4a/4b	eNOS	4a4a	ACS/controls (477/537)	Endothelial/ vascular function	2.5 (1.1–5.4) for ACS risk, 3.6 (1.2–11.5) for MI risk
Antoniades et al. [31, 32,35,37]	eNOS/894G>T	eNOS	GT,TT	MI/controls (228/519)	Endothelial/ vascular function	1.192 (1.131–3.485)
Keavney B et al. [33]	ACE/I/D	ACE	No difference	MI (4629)	Endothelial/ vascular function	1.10 (1.00–1.21)
Keavney B et al. [33]	AT1R/1166A>C	AT1R	AA/AC in individuals with ACE DD	MI (4629)	Endothelial/ vascular function	3.95 with ACE DDgenotype

Antoniades et al. [31, 32,35,37]	ATG/ rs8007267G>A, rs3783641A>T, rs10483639C>G	GTP-cyclohydrolase I	X haplotype	CAD (347)	Endothelial/ vascular function	Not studied
Santoso et al. [36]	GPIa/807C>T	GPIa	CT,TT	MI<62 years / MI<49 years (1057/223)	Thrombosis/ fibrinolysis	0.57 for pts<62 yrs old, 2.61 for pts<49 yrs old
Antoniades et al. [31, 32,35,37]	GPIa/807C>T	GPIa	CT,TT	MI/controls (219/389)	Thrombosis/ fibrinolysis	2.296
Reiner et al. [38]	GPIIb/Ser 843	GPIIb	Ser 843	MI/controls (68/346)	Thrombosis/ fibrinolysis	1.85 for women under 42 yrs old
Bojesen et al. [39]	GPIIIa/Pl(A2)/Pl(A2), Pl(A1)/Pl(A2), Pl(A1)/Pl(A1)	GPIIIa	Pl(A2)/Pl(A2)	Patients, prospective (9149)	Thrombosis/ fibrinolysis	RR = 3.6 for patients with MI<40 yrs old
Topol et al. [40]	Thrombospondin 4/ Ala387Pro	Thrombospondin	AlaPro, ProPro	Families (men <45 years and/ or women <50 years MI) (398)	Thrombosis/ fibrinolysis	1.89
Ding et al. [43]	PAI-1/4G/4G haplotype	PAI-1	4G/5G, 4G/4G	70–79 years, MI (177)	Thrombosis/ fibrinolysis	No linkage with MI
Klerk et al. [44]	MTHFR/677C>T	MTHFR	TT	CHD/controls (11,162/12,758)	Thrombosis/ fibrinolysis	1.16 (1.05–1.28)
Mälarstig et al. [46,51]	IL-6/572G>C, −598C>T	IL-6	CG	ACS/controls (3027/447)	Inflammation	−572 CG genotype was predictive for increased IL-6 levels in patients with a subsequent cardiovascular event, 2.3 (1.1–4.3)

Continued

TABLE 7.1 Polymorphisms Associated With Lipid Metabolism, Endothelial Function, Thrombosis/Fibrinolysis, and Inflammation—cont'd

Study	Gene/ Polymorphism	Protein Encoded	Risk Genotype	Patients (Number)	Associated With	OR (CI) for ACS
Ozaki et al. [47]	LTA-a	LTA-a	Thr26Asn in coding region, SNP in intron-1	MI/controls (1.133/1.006)	Inflammation	1.78
Bis et al. [48]	IL-18/SNP rs2043055	IL-18	IL-18 haplotype	MI controls (856/2688)	Inflammation	0.74 (0.64–0.87)
Park et al. [49]	PAPP-A	PAPP-A	IVS6+95C allele	MI/controls (170/170)	Inflammation	2.13 (1.12–4.07)
Simeoni et al. [50]	RANTES/403G>A	RANTES	AG,AA	CAD/controls (2694/530)	Inflammation	1.36 (1.08–1.75)
Mälarstig et al. [46,51]	CD40LG/3459A>G	CD40L	AA,GG,AG	ACS (2359)	Inflammation	2.50
Ameziane et al. [52]	TLR4/Asp299Gly	TLR4	GlyGly,AspGly	ACS/controls (183/216)	Inflammation	0.41 (0.18–0.98)
Horne et al. [53]	MMP-9	MMP-9	CT/RQ	MI/controls (1967/3455)	Inflammation	1.25

ApoE, Apolipoprotein E; *OLR1*, oxidized LDL receptor 1; *CETP*, cholesterol ester transfer protein; *LPL*, lipoprotein lipase; *MTHFR*, methylenetetrahydrofolate reductase; *ACS*, acute coronary syndrome; *eNOS*, endothelial NO synthase; *ACE*, angiotensin converting enzyme; *AT1R*, angiotensin II type I receptor; *MI*, myocardial infarction; *CAD*, coronary artery disease; *GPIa*, Glucoprotein Ia; *GPIIb*, glucoprotein IIb; *GPIIIa*, glucoprotein IIIa; *PAI-1*, plasminogen activator inhibitor 1; *IL-6*, interleukin-6; *LTA-a*, lymphotoxin-a; *IL-18*, interleukin-18; *PAPP-A*, pregnancy-associated plasma protein-A; *RANTES*, regulated upon activation, normal T-cell expressed and secreted; *CD40LG*, CD40 ligand gene; *TLR-4*, toll-like receptor 4; *MMP-9*, metalloproteinase-9.

association of APOE alleles with MI was analyzed in 1817 patients admitted for first premature MI. In patients without hypercholesterolemia, the APOE allele ε4 was highly predictive for the presence of premature MI [20]. A recent meta-analysis of lipid levels and coronary risk studies replicated the results of the aforementioned study. Compared with the ε3 carriers the ε4 carriers had 6% greater risk of CAD (OR 1.06, 95% CI, 0.99–1.13) [21].

The oxidized LDL receptor 1 (OLR1) gene encodes a receptor involved in endocytosis at the cell surface for oxidized LDL. In patients with a history of MI, Tatsuguchi et al. [22] found significantly higher frequency of the 501G>C (38.2%) compared to controls (17.6%). The OR for the risk of MI with the polymorphism was 2.89 [22].

Lipoprotein lipase (LPL) is an enzyme responsible for hydrolyzing triglycerides, resulting in the release of fatty acids and glycerol and the production of high-density lipoprotein (HDL) cholesterol. The most common polymorphism of LPL in the general population appears to be the Ser447Ter with a heterozygote frequency of 17–22%. According to a meta-analysis it is associated with decreased triglyceride and increased HDL levels in carriers, while the Gly-188Glu polymorphism with heterozygote frequencies of 0.04–0.2% augments the risk of coronary heart disease [23].

It is well known that cholesterol ester transfer protein (CETP) mediates the uptake of cholesterol from peripheral tissues to the liver, transfers the cholesteryl ester to and receives triglycerides from very low-density lipoprotein (VLDL) and LDL. Molecular variations in the gene coding for the CETP such as the TaqIB polymorphism are associated with higher plasma HDL concentration. In a prospective cohort of 14,916 apparently healthy men, the TaqIB polymorphism was determined among 384 individuals who developed a first MI and 384 controls. Carriers of the B2 allele of the TaqIB in the CETP gene had higher HDL concentrations but did not have lower risk of MI (relative risk 0.95, 95% CI 0.54–1.66) [24]. However, a later meta-analysis, associated B2B2 genotype with both high HDL concentrations and lower risk of CAD (OR 0.78, $P=0.008$, 0.66–0.93) [25].

Polymorphisms Involved in Endothelial and Vascular Function (Table 7.1)

At least three types of endothelial nitric oxide synthase (eNOS) produce nitric oxide (NO), which regulates vascular tone. With the exception of NO production, nitric oxide synthase is a key enzyme for cardiovascular homeostasis. One of nitric oxide's physiologic roles is to help prevent atherosclerotic changes in the vasculature by inhibiting smooth muscle proliferation and platelet adhesion and aggregation. The 393 bp allele (eNOS4a) of eNOS has been studied in African Americans and has been more common amongst MI patients than controls (36% vs 26%, $P=0.01$) [26]. In comparison to control subjects, individuals heterozygous for the 393 allele had an odds ratio for MI of 1.8 ($P=0.03$), while that of homozygotes was

2.2 (P=0.07). In the same study, significant correlation between the presence of at least one GPIIIa PlA2 allele (which codes for a glycoprotein receptor) and at least one eNOS 393 allele (eNOS4a) with an increased OR for MI (OR 4.3, P=0.01) has been observed [26]. In a Caucasian study, eNOS4a homozygotes who were current or former smokers had a more frequent history of MI than did smokers with other genotypes [27]. In another Japanese case–control study, eNOS 393 allele (eNOS4a) augments MI risk independently of smoking status and other risk factors [28]. Fatini et al. also proved that the 4a4a genotype causes a predisposition to ACS (OR 2.5) and particularly to MI (OR 3.6) [29]. A meta-analysis of 23,028 subjects involved in 26 studies inquired about the role of Glu298Asp and intron-4 (eNOS 4a, 4b) polymorphisms. Subjects homozygous for the eNOS 4a allele were at an increased risk for ischemic heart disease [30].

We have shown that the genetic polymorphism 894G>T on eNOS is associated with increased MI risk. In a population of 228 patients with a first event of premature MI (<49 years old) and 519 controls the risk for MI in 894TT homozygotes was 1.992 (95% confidence interval, 1.131–3.485, P<0.05) versus GG+GT individuals [31,32].

Furthermore, the deletion/deletion (DD) intron 16 genotype of angiotensin converting enzyme (ACE) has been linked to an increased risk of MI in smaller studies, but larger studies have not supported this association. A multicenter case–control study examining a large group of 4629 MI cases found that the angiotensin-converting enzyme (ACE) DD genotype was associated with a risk ratio for MI in the range of 1.0–1.1, suggesting that the ACE gene is not associated with an increased susceptibility to MI [33]. In individuals with the ACE DD genotype the 1166A>C polymorphism (A>C transversion) of the AT1R (angiotensin II type-1 receptor) gene conferred an increased risk of MI (OR 1.52 for A/C heterozygotes, 3.95 for C/C homozygotes) [33].

All NOS isoforms along with other biosynthetic pathway enzymes are reported to share tetrahydrobiopterin (BH4) as cofactor. Reduced BH4 availability seems to impair eNOS activity and increase vascular superoxide production [34]. The rate-limiting step of BH4 biosynthesis is catalyzed by guanosine triphosphate (GTP) cyclohydrolase I (GCH1). Recently, Antoniades et al. deciphered the effects of GCH1 gene on BH4 bioavailability and vascular function in a candidate gene association study of CAD patients. Among the three haplotypes studied, the "X"-designated haplotype was an independent predictor of reduced BH4 levels in both plasma and vascular tissue and determinant of eNOS coupling, vascular redox state, and endothelial function in patients with CAD [35].

Polymorphisms Involved in Thrombosis/Fibrinolysis Cascades (Table 7.1)

The glycoprotein (GP) Ia–IIa complex mediates adherence of platelets to collagen. The T807 allele was strongly associated with nonfatal MI among individuals younger than 62, in both heterozygotes and homozygotes (OR 1.57,

$P=0.004$). The highest risk was observed in individuals under 49 years old (OR 2.61, $P=0.009$) [36]. In the same study of 2237 men undergoing coronary angiography, the C807 was not associated with MI risk [34]. We have previously indicated in a study of 219 patients with premature MI and 389 controls that the risk for MI in 807TT homozygotes was 2.296 (95% confidence interval, 1.187–4.440, $P<0.05$) versus CC+CT [37].

The 843 allele of the glycoprotein IIb Ser843 polymorphism has been found to increase MI risk in young women 18–42 years old (OR 1.85) [38]. A GPIIIa polymorphism, known as the PlA1/PlA2 or HPA-1a/HPA-1b polymorphism enhances fibrinogen binding to platelets and increases platelet aggregation in young (<40 years) men. P1A2/P1A2 homozygotes are associated with a four-fold increase in risk of MI [39].

Studies indicate that the Ala387Pro variant (Pro allele) of thrombospondin-4 is most strongly associated with an increased risk of MI. Carriers of the Pro allele had an OR for MI of 1.89 ($P=0.002$) in a study of 350 families with premature MI [40]. A thrombospondin-2 variant, on the other hand, appeared to be associated with a decreased risk of MI (OR 0.31, $P=0.0018$) [40].

Plasminogen activator inhibitor-1 (PAI-1) plays an important role in regulating intravascular thrombosis and thrombolysis. The PAI-1 4G allele of the common insertion/deletion polymorphism (4G/5G) of PAI-1 is associated with increased levels of circulating PAI-1, as it allows only a transcriptional activator but not a repressor gene to bind to the promoter. In an initial study the 4G allele frequency was significantly higher in patients who had an MI before age 45 than in the general population (0.63 vs 0.53) [41]. However, in a subsequent study of 1353 patients undergoing coronary angiography there was no association between having the 4G allele MI risk [42]. Recently, in a prospective analysis of 2995 community-based participants (41% blacks and 51% women), 177 of whom suffered from MI, the 4G allele was associated with higher PAI-1 levels but not with elevated risk of MI [43].

Elevated homocysteine levels promote activation of thrombotic cascades and increase cardiovascular risk. An important enzyme involved in the remethylation of homocysteine to methionine is methylene-tetrahydrofolate reductase (MTHFR). Reduced activity of this enzyme will result in an elevation in plasma homocysteine levels. A mutation identified in the MTHFR gene (677C>T) results in reduced enzyme activity in folate-deficient individuals. About 12% of Caucasians are homozygous for this mutation (TT). In a meta-analysis of 11,162 CHD subjects and 12,758 controls from 40 studies, the MTHFR TT genotype was associated with an increased risk of CHD (OR 1.16, 1.05–1.28) under low-folate conditions [44].

Polymorphisms of Genes Involved in Inflammation (Table 7.1)

The association between two interleukin 6 (IL-6) promoter polymorphisms, the IL-6-572G>C and IL-6-174G>C polymorphisms, and MI was investigated in the study Hypercoagulability and Impaired Fibrinolytic Function MECHanisms Predisposing to Myocardial Infarction [45]. The North American carriers of −74C allele exhibited lower risk of acute MI (AMI) compared to −174GG homozygotes, while no association with MI was observed for the −572G>C polymorphism (OR 0.53, 0.32–0.86). The south European homozygotes had no increased risk for MI [45]. Similarly, in a recent study no association between the −572G>C polymorphism has been documented [46].

Homozygotes for two polymorphisms in LTA gene (lymphotoxin-a) have been found to face a significant increased risk for MI (OR 1.78, $P=0.00000033$; 1133 affected individuals versus 1006 controls) [47]. In a recent case–control study of genetic variation among eight inflammation-related genes (interleukin (IL)-1β, IL-6, C-reactive protein (CRP)), IL-10, IL-18, and the tumor necrosis factor (TNF) superfamily [lymphotoxin(LT)-α, TNF-α, LT-β], only one gene (IL-18) was associated with MI risk (OR, 0.74; 95% CI, 0.64–0.87) [48]. In another study of 170 patients with MI, pregnancy-associated plasma protein-A (PAPP-A, a zinc-binding matrix metalloproteinase that is known to be abundantly expressed in ruptured plaques) IVS6+95C allele is an independent risk factor for MI (OR, 2.13, 95% confidence interval, 1.12–4.07, $P=0.022$) [49].

Another important inflammatory molecule, regulated upon activation, normal T-cell expressed and secreted (RANTES), mediates T-cell and monocyte recruitment and has been observed in T-cells in human atherosclerotic plaques. In a case–control study of 2694 patients with CAD (documented with coronary angiography) and 530 controls, RANTES A-403 was associated with increased ACS prevalence (OR 1.36, 95% CI 1.08–1.71, $P=0.0073$) [50].

CD40 is expressed on a variety of immune cells within atherosclerotic lesions including endothelial cells, smooth muscle cells, and monocytes/macrophages, whereas CD40L is largely expressed on CD4 + T-cells and activated platelets. Elevated sCD40L levels at hospital admission have been associated with increased risk of ACS (OR=2.50, $P<0.001$). The −3459A>G polymorphism in the CD40LG gene was associated with increased level of CD40L but not increased risk of myocardial infarction [51]. We have shown that the 807T allele of the 807C>T polymorphism of platelet glycoprotein Ia was associated with higher sCD40L levels compared with 807CC ($P<0.05$) [37].

Toll-like receptor 4 (TLR4) is a transmembrane receptor that is involved in mediating inflammatory responses to bacterial endotoxin and other ligands. The TLR4 Asp299Gly polymorphism was investigated in a case–control study of 183 patients with acute coronary syndromes and 216 controls and was associated decreased risk of acute coronary events independently of standard coronary risk factors (OR 0.41, 95% CI 0.18–0.95, $P=0.037$) [52].

Matrix metalloproteinases (MMPs) mediate degradation of the extracellular matrix of coronary plaque, hence they render the plaque unstable. Matrix metalloproteinases are suppressed by tissue inhibitors of metalloproteinases. In the largest case–control study the MMP-9 haplotype CT/RQ was associated with greater MI risk compared to CC/RR (odds OR 1.25, $P=0.007$) [53].

GENOME-WIDE LINKAGE AND ASSOCIATION STUDIES: LEAPS TOWARD INDIVIDUALIZED CARDIOVASCULAR MEDICINE?

So far, the nine genome-wide linkage studies have identified 10 regions of linkage on loci 1p36 [54], 2p11 [54,55], 2q21 [56], 2q36 [57], 3q13 [58], 12q22 [59], 13q12 [59], 14 [60], 15q26 [61], 16q13 [62], 17 [63], and Xq25 [57], and only two loci replicated from one study to another 2p11 [54,55,64], 3q13 [64], and five genes have been thus far directly associated with CAD: (1) and (2) The haplotype of ALOX5AP-5-lipooxygenase activating protein (FLAP) under the linkage peak at 13q12-13 [59], and the haplotype of leukotriene A4 hydrolase (LTA4H) after fine mapping analysis for the 12q22 linkage peak [65] are two genes in the pathway of leukotriene B4 associated with MI. However, the results for ALOXAP haplotypes have not been replicated in a very recent association study [66]; (3) The myocyte enhancer factor 2a (MEF2A) on 15q26 was identified in a family with autosomal dominant inheritance pattern of MI or CAD [61]. This linkage has not been replicated in a recent study including multiple families with apparently Mendelian inheritance of MI or CAD [67]; (4) Vesicle-associated membrane protein 8 (VAMP8) gene on a chromosome 2 region, which has been identified in several linkage study peaks, is recently associated with MI in a population with early onset MI [68]. The protein encoded is required for platelet-dense granule secretion and platelet activation; (5) The KALiRiN gene on 3q13 locus, belonging to the Rho GTPase-signaling pathway, has been proposed to be significantly associated with early onset CAD cases among individuals of European ancestry [69]. Whether this gene is functionally associated with MI or not remains to be seen. The lack of strong evidence regarding gene contribution to CAD observed in linkage studies could also be attributed to the use of no more than 400 microsatellite markers per study, which means searching disease-causing genes interspersed between 1 million base pairs [70].

Genome-Wide Association Studies and 9p21 Locus (Table 7.2)

Some of the 10 million SNPs that exhibit common variation between individuals in a population affect the risk of complex diseases such as ACS. The complex genetic variation of these diseases motivates whole genome association studies with SNP markers. Currently, the haplotype map (International HapMap) project created a public resource of common SNPs throughout the human genome sequence and provided the LD structure of genomes based on these SNPs [71]. The existing, continuously updated databases provide over 4 million validated SNPs [72]. However, for large eukaryotic genomes such as the human, even recently developed high-throughput sequencing technologies only allow genome-wide sequence analysis of a small number of individuals. Resequencing the genome of many individuals for which there is a reference genome allows investigation of the relationship between sequence variation and normal or disease phenotypes. Novel sequencing approaches involve sequencing by synthesis on clonal single-molecule templates. Systems applying synthesis on

TABLE 7.2 Most Important Genome-Wide Association Studies of 9p21 With Coronary Artery Disease and/or Myocardial Infarction

Study	Study Design	Patients/ Controls	SNPs on Array	Chromosome Loci/Haplotypes	Risk of CAD for Risk Allele 0R (95% CI)
McPherson et al. [73]	Case–control genome-wide association study	322/312	100.000	9p21/rs10757274 9p21/rs2383206	1.18 (1.02–1.37) 1.26 (1.09–1.46)
Helgadottir et al. [74]	Case–control genome-wide association study	1607/6728	300.000	9p21/rs10757278 9p21/rs2383207	1.26 (1.16–1.36) and 2.02 (1.72–2.36) for early onset CAD
Wellcome Trust Case–Control Consortium [75]	Case–control genome-wide association study	1988/3004	500.000	9p21/rs1333049	1.47 (1.27–1.70) heterozygous
Samani et al. [76]	Case–control genome-wide association study	875/1644	500.000	9p21.3/rs1333049	1.37 (1.26–1.48)
Schunkert et al. [77]	Replication study	4645/5177		9p21.3/rs1333049	1.29 (1.22–1.37)
Schunkert et al. [77]	Meta-analysis of all the above studies	12004/28949		9p21.3/rs1333049	1.24 (1.20–1.29)
Helgadottir et al. [88]	Association study of 9p21 locus with abdominal aneurysm, intracranial aneurysm and myocardial infarction	1607 14278		9p21.3/rs10757278	1.25 (1.14–1.37)
Broadbent et al. [86]	PROCARDIS consortium replication study of 9p21 haplotypes and its association with risk factors	4251/4443		9p21.3/rs2891168	1.29 (1.20–1.38)

Assimes et al. [78]	ADVANCE multiethnic case–control replication study	Non blacks 1414/1424 blacks 96/325		9p21.3/ (rs10757274, rs2383206, and rs10757278 – G haplotype)	1.23 (0.99–1.52 for whites) 1.10 (0.66–1.83 for blacks-no significant association)
Abdullah et al. [90]	Case–control study of 4 new 9p21 SNPs associated with premature CAD in American Caucasians	310/560		9p21/(rs10757274, rs2383206, rs2383207, and rs10757278)	1.78 (1.45–2.18), 1.67 (1.36–2.03), 1.72 (1.41–2.12), 1.76 (1.43–2.14), respectively
Talmud et al. [91]	Prospective study of clinical utility of 9p21 haplotypes	270/2742 (followed for 14 years		9p21.3/rs10757274, rs2383206	HR 1.47 (1.08–1.99)–improved reclassification of cardiovascular risk
Samani et al. [89]	association study of 9p21 haplotypes with early markers of atherosclerosis	2277 young healthy individuals/1295 older healthy individuals		9p21.3/rs1333049	No significant association with flow-mediated dilation and carotid intima-media thickness
Shen et al. [80]	Replication study of four SNPs in Asian South Korean population	611/294		9p21.3/rs10757274 and rs2383206 for CAD and rs2383207 and rs10757278 with MI-GGGG haplotype	1.76 (1.54–2.37)
Shen et al. [92]	Replication study of 4 SNPs in Mediterranean population	416 with MI/318		9p21.3/rs10757274 and rs2383206 for CAD and rs2383207 and rs10757278 with MI- GGGG haplotype	1.58 (1.41–1.79) association limited to those with MI or CAD family history

CAD, Coronary artery disease; *SNPs*, single nucleotide polymorphisms.

arrays sequencing are capable of studying heterozygous SNP alleles in diploid DNA as in the case of CAD association studies. Direct sequencing from single DNA molecules is a promising approach providing high data density with small amounts of sample.

Currently the strongest and most robust susceptibility locus identified by four genome-wide association studies for CAD is located on chromosome 9p21 [73–76]. A recent meta-analysis [77] of seven European populations and the four populations that participated in previous genome-wide association studies of CAD [73–76] provided evidence for consistent associations between the SNP rs1333049 on locus 9p21.3 and CAD and MI. The OR per copy of the risk allele was 1.29 (95% confidence interval, 1.22 to 1.37; $P<0.0001$). The most likely inheritance model is the autosomal additive model ($P<0.38$). Ethnic heterogeneity might be an issue in association studies of 9p21. The multiethnic ADVANCE study [78] and two case–control studies have obtained genetic risks similar to those found in European and North American populations in other studies of 9p21 haplotypes [79,80]. In contrast, there is a lack of association in African American subjects as concluded in two of the aforementioned case–control studies [73,78]. The variety of results among different ethnic groups could be attributed to limited samples, interactions with environmental factors, deficient LD among the examined SNPs on 9p21 locus, and interactions with other genes whose allele frequencies vary across ethnic groups.

Some [81–83] but not all [84,85] genome-wide association studies of type 2 diabetes have also identified some susceptibility SNP markers within 9p21 locus. The PROCARDIS consortium conducted an association study of the 9p21 SNPs previously associated with CAD and type 2 diabetes in four European populations in order to evaluate the contribution of conventional risk factors (age, gender, obesity, hypertension, smoking, diabetes, and onset age of CAD) to susceptibility for CAD [86]. No association of studied haplotypes with risk factor variation has been reported. SNPs showing the strongest association with CAD did not map within an annotated gene sequence, leading to the nomination of neighboring genes (CDKN2A, CDNK2B, and MTAP) as candidate susceptibility genes. The high-risk haplotypes of the PROCARDIS consortium study have been in LD with ANRIL, a large antisense noncoding RNA gene [83], shown to be expressed in tissues and cell types affected by atherosclerosis. The functions of this gene may be regulated by CDKN2A and CKKN2B [87], therefore this novel mechanism leading to CAD phenotype independently of risk factors warrants more study. Indicative of the forthcoming breakthrough in the exploration of CAD genetic background is the fact that 9p21 haplotypes are associated with various vascular diseases such as CAD, abdominal aortic aneurysm, and intracranial aneurysm [88] but not with functional assays of atherosclerosis (flow-mediated dilation and carotid intima media thickness) [89]. It is possible that 9p21 locus influences CAD risk through a mechanism not associated with early atherosclerotic changes, contrary to what conventional risk factors do.

Despite recent advances in cardiovascular genetics, for the vast majority of mapped loci it has been a challenge to move from genomic localization to biologic mechanism. Genome-wide association studies provide gene regions and not necessarily specific causal variants or causal genes. For each SNP and locus mapped by genome-wide association studies, we can generally conclude that within ~100,000 bases of the locus, there exists a causal gene. Ideally, at each discovered locus, we need to understand: (1) the causal variant; (2) the causal gene; (3) the mechanism by which the variant affects the gene; and (4) the mechanism by which the gene affects the phenotype. Several features of genetic mapping using common variants have contributed to the difficulties. Linkage disequilibrium and weak effect size of common variants have made it difficult to identify causal variants. Mapped SNP variants have weak or modest effect for a dichotomous trait, and these weak effects mean that in order to statistically distinguish between two correlated variants, the sample sizes need to be inordinately large. Nearly all of the mapped SNP variants are noncoding, and our ability to interpret the noncoding portion of the genome remains limited. Mechanisms by which a noncoding variant may affect a nearby gene include affecting a transcription factor or microRNA binding site and regulating the local chromatin state.

Mendelian Randomization Studies

Both causal and noncausal biomarkers may be helpful in terms of predicting risk for future disease. However, only a causal biomarker may be appropriate as a target of therapy. The cost and logistical complexity of following a very large number of young asymptomatic people over several decades within the context of a clinical trial may be impractical. However, in the absence of a randomized trial, it may still be possible to estimate the effect of random allocation to long-term exposure to a risk factor on the risk of CAD by appealing to the principle of Mendelian randomization or "natural randomized study." Mendelian randomization is the random assignment of an individual's genotype from his or her parental genotypes that occurs before conception can be exploited to make causal inferences in etiological epidemiology. The use of Mendelian randomization has the potential to avoid some of the limitations of observational epidemiology (confounding, reverse causality, regression dilution bias) and randomized controlled studies (generalizability, unfeasible for some exposures, expensive) for making causal inferences. However, the approach has a number of potential limitations, including lack of suitable genetic variants, unreliable associations, population stratification and the need for large sample sizes, and some potential problems with binary outcomes [90]. Mendelian randomization approach has been utilized to determine the magnitude of causal relationships between CRP and carotid intima media thickness [91], fibrinogen and coronary heart disease [92], and prolonged exposure to elevated LDL and hypertension and CAD [93].

CONCLUSIONS

The heterogeneity of results and lack of replication are major drawbacks of candidate gene studies. In addition the inconclusive results of genetic linkage and association studies of acute coronary syndrome patients indicate family history and classic risk factor control as the safest tools to create patient-specific risk profiles for acute coronary syndromes. Definitely, whole genome association studies unveiled novel aspects of the complex pathophysiology of CAD. However, only rarely will the SNPs originally indicating a disease susceptibility locus prove to be functional. Exhaustive resequencing will thus be a prerequisite to identify potentially functional candidates for subsequent studies and to address the issue of rare variants with high penetrance. In the era of individualized medicine, for any newly identified candidate gene, the exact biological pathways in which the gene product takes part need to be established.

GLOSSARY

Alleles Variant forms of the DNA sequence at a specified locus.

Candidate genes Genes suspected of involvement in the disease process.

Genomics The investigation of the structure and function of large sections of the genome simultaneously.

Haplotype A combination of alleles that are located at closely linked loci and tend to be inherited together.

Linkage Two genetic loci are linked if they are transmitted together from parent to offspring more often than expected under independent inheritance.

Linkage analysis A statistical method that is conducted in families with several affected members, aiming to locate a gene causally related to a disease by identifying genetic markers of known chromosomal location that are in LD with the trait of interest.

Linkage disequilibrium (LD) Two loci are in LD if across the population as a whole they are found together on the same haplotype more often than expected.

Microarrays Novel techniques designed to study interactions and expression regulation of vast numbers of genes. Small quantities of functional DNA are applied on glass slides. The DNA probes to be studied are fluorescently labeled and allowed to bind to cDNA strands on the slides. The slides are put into a scanning microscope that can measure the brightness of each fluorescent dot; brightness reveals how much of a specific DNA fragment is present, which is a quantitative indicator of its expression.

Microsatellite A variable locus characterized by different numbers of repetitions (sequences of two to five nucleotides) of the same motif in tandem.

Polymorphism Any variation in the sequence of DNA among individuals with minor allele frequency $<1\%$.

Resequencing Precisely determining the sequence of nucleotides in the DNA of a sample of individuals. Resequencing is used to identify single-nucleotide polymorphisms.

Whole genome association study A study that investigates statistical associations between a phenotype and a very large number of genetic markers (SNPs). These studies do not rely on a priori knowledge and may lead to the discovery of new causes of disease. They do not resequence the entire genome but type a large set of markers distributed throughout the genome.

REFERENCES

[1] Greenland P, et al. Major risk factors as antecedents of fatal and nonfatal coronary heart disease events. JAMA 2003;290:891–7.

[2] Khot UN, et al. Prevalence of conventional risk factors in patients with coronary heart disease. JAMA 2003;290:898–904.

[3] Yusuf S, et al. Effect of potentially modifiable risk factors associated with myocardial infarction in 52 countries (the INTERHEART study): case-control study. Lancet 2004;364:937–52.

[4] Shea S, et al. Family history as an independent risk factor for coronary artery disease. J Am Coll Cardiol 1984;4:793–801.

[5] Friedlander Y, et al. Family history as a risk factor for primary cardiac arrest. Circulation 1998;97:155–60.

[6] Leander K, et al. Family history of coronary heart disease, a strong risk factor for myocardial infarction interacting with other cardiovascular risk factors: results from the Stockholm Heart Epidemiology Program (SHEP). Epidemiology 2001;12:215–21.

[7] Sesso HD, et al. Maternal and paternal history of myocardial infarction and risk of cardiovascular disease in men and women. Circulation 2001;104:393–8.

[8] Lloyd-Jones DM, et al. Parental cardiovascular disease as a risk factor for cardiovascular disease in middle-aged adults: a prospective study of parents and offspring. JAMA 2004;291:2204–11.

[9] Murabito JM, et al. Sibling cardiovascular disease as a risk factor for cardiovascular disease in middle-aged adults. JAMA 2005;294:3117–23.

[10] Jorde LB, et al. Medical genetics, 3rd ed. St. Louis: Mosby; 2003.

[11] Marenberg ME, et al. Genetic susceptibility to death from coronary heart disease in a study of twins. N Engl J Med 1994;330:1041–6.

[12] Zdravkovic S, et al. Heritability of death from coronary heart disease: a 36-year follow-up of 20966 Swedish twins. J Intern Med 2002;252:247–54.

[13] Reich DE, et al. On the allelic spectrum of human disease. Trends Genet 2001;17:502–10.

[14] Lohmueller KE, et al. Meta-analysis of genetic association studies support a contribution of common variants to susceptibility to common diseases. Nat Genet 2003;33:177–82.

[15] Bodmer W, et al. Common and rare variants in multifactorial susceptibility to common diseases. Nat Genet 2008;40:695–701.

[16] Pritchard JK, et al. Are rare variants responsible for susceptibility to complex diseases. Am J Hum Genet 2001;69:124–37.

[17] Goring HH, et al. Large upward bias in estimation of locus-specific effects from genomewide scans. Am J Hum Genet 2001;69:1357–69.

[18] Hattersley AT, et al. What makes a good genetic association study? Lancet 2005;366:1315–23.

[19] Morgan TM, et al. Nonvalidation of reported genetic risk factors for acute coronary syndrome in a large-scale replication study. JAMA 2007;297:1551–61.

[20] Schmitz F, et al. Robust association of the APOE epsilon4 allele with premature myocardial infarction especially in patients without hypercholesterolaemia: the Aachen study. Eur J Clin Invest 2007;37(2):106–8.

[21] Bennett AM. Association of apolipoprotein E genotypes with lipid levels and coronary risk. JAMA 2007;298(11):1300–11.

[22] Tatsuguchi M, et al. Oxidized LDL receptor gene (OLR1) is associated with the risk of myocardial infarction. Biochem Biophys Res Commun 2003;303:247–50.

[23] Wittrup HH, et al. Lipoprotein lipase mutations, plasma lipids and lipoproteins, and risk of ischemic heart disease: a meta-analysis. Circulation 1999;99:2901–7.

[24] Liu S, et al. A prospective study of TaqIB polymorphism in the gene coding for cholesteryl ester transfer protein and risk of myocardial infarction in middle-aged men. Atherosclerosis 2002;161(2):469–74.
[25] Boekholdt SM, et al. Cholesteryl ester transfer protein TaqIB variant, high-density lipoprotein cholesterol levels, cardiovascular risk, and efficacy of pravastatin treatment: individual patient meta-analysis of 13,677 subjects. Circulation 2005;111:278–87.
[26] Hooper WC, et al. The relationship between polymorphisms in the endothelial cell nitric oxide synthase gene and the platelet GPIIIa gene with myocardial infarction and venous thromboembolism in African Americans. Chest 1999;116:880–6.
[27] Wang XL, et al. A smoking-dependent risk of coronary artery disease associated with a polymorphism of the endothelial nitric oxide synthase gene. Nat Med 1996;2:41–5.
[28] Ichihara S, et al. Association of a polymorphism of the endothelial constitutive nitric oxide synthase gene with myocardial infarction in the Japanese population. Am J Cardiol 1998;81:83–6.
[29] Fatini C, et al. Influence of endothelial nitric oxide synthase gene polymorphisms (G894T, 4a4b, T-786C) and hyperhomocysteinemia on the predisposition to acute coronary syndromes. Am Heart J 2004;147(3):516–21.
[30] Casas JP, et al. Endothelial nitric oxide synthase genotype and ischemic heart disease: meta-analysis of 26 studies involving 23028 subjects. Circulation 2004;109:1359–65.
[31] Antoniades C, et al. Effects of endothelial nitric oxide synthase gene polymorphisms on oxidative stress, inflammatory status, and coronary atherosclerosis: an example of a transient phenotype. J Am Coll Cardiol 2006;48(6):1166–74.
[32] Antoniades C, et al. Genetic polymorphisms G894T on the eNOS gene is associated with endothelial function and vWF levels in premature myocardial infarction survivors. Int J Cardiol 2006;107(1):95–100.
[33] Keavney B, et al. Large-scale test of hypothesized associations between the angiotensin-converting-enzyme insertion/deletion polymorphism and myocardial infarction in about 5000 cases and 6000 controls. Lancet 2000;355:434–42.
[34] Channon KM. Tetrahydrobiopterin: regulator of endothelial nitric oxide synthase in vascular disease. Trends Cardiovasc Med 2004;14:323–7.
[35] Antoniades C, et al. GCH1 haplotype determines vascular and plasma biopterin availability in coronary artery disease: effects on vascular superoxide production and endothelial function. J Am Coll Cardiol 2008;52:158–65.
[36] Santoso S, et al. Association of the platelet glycoprotein Ia C807T gene polymorphism with nonfatal myocardial infarction in younger patients. Blood 1999;93:2449–53.
[37] Antoniades C, et al. Genetic polymorphisms of platelet glycoprotein Ia and the risk for premature myocardial infarction: effects on the release of sCD40L during the acute phase of premature myocardial infarction. J Am Coll Cardiol 2006;47(10):1959–66.
[38] Reiner AP, et al. Platelet glycoprotein IIb polymorphism, traditional risk factors and non-fatal myocardial infarction in young women. Br J Haematol 2001;112:632–6.
[39] Bojesen SE, et al. Platelet glycoprotein IIb/IIIa P1A1/P1A2 homozygosity associated with risk of ischemic cardiovascular disease and myocardial infarction in young men: the Copenhagen City Heart Study. J Am Coll Cardiol 2003;42:661–7.
[40] Topol EJ, et al. Single nucleotide polymorphisms in multiple novel thrombospondin genes may be associated with familial premature myocardial infarction. Circulation 2001;104:2641–4.
[41] Eriksson P, et al. Allele-specific increase in basal transcription of the plasminogen activator inhibitor 1 gene is associated with myocardial infarction. Proc Natl Acad Sci USA 1995;92:1851–5.

[42] Anderson JL, et al. Lack of association of a common polymorphism of the plasminogen activator inhibitor-1 gene with coronary artery disease and myocardial infarction. J Am Coll Cardiol 1999;34:1778–83.
[43] Ding J, et al. Plasminogen activator inhibitor type 1 gene polymorphisms and haplotypes are associated with plasma plasminogen activator inhibitor type 1 levels but not with myocardial infarction or stroke. Am Heart J 2006;152(6):1109–15.
[44] Klerk M, et al. MTHFR 677C>T polymorphism and risk of coronary heart disease: a meta-analysis. JAMA 2002;288:2023–31.
[45] Kelberman D, et al. Effect of Interleukin-6 promoter polymorphisms in survivors of myocardial infarction and matched controls in the North and South of Europe. The HIFMECH Study. Thromb Haemost 2004;92(5):1122–8.
[46] Mälarstig A, et al. Genetic variation in the interleukin-6 gene in relation to risk and outcomes in acute coronary syndrome. Thromb Res 2007;119(4):467–73.
[47] Ozaki K, et al. Functional SNPs in the lymphotoxin-alpha gene that are associated with susceptibility to myocardial infarction. Nat Genet 2002;32(4):650–4.
[48] Bis JC, et al. Variation in inflammation-related genes and risk of incident nonfatal myocardial infarction or ischemic stroke. Atherosclerosis 2008;198(1):166–73.
[49] Park S, et al. Genetic polymorphism in the pregnancy-associated plasma protein-A associated with acute myocardial infarction. Coron Artery Dis 2007;18:417–22.
[50] Simeoni E, et al. Association of RANTES G-403A gene polymorphism with increased risk of coronary arteriosclerosis. Eur Heart J 2004;25(16):1438–46.
[51] Mälarstig A, et al. Soluble CD40L levels are regulated by the −3459A>G polymorphism and predict myocardial infarction and the efficacy of antithrombotic treatment in non-ST elevation acute coronary syndrome. Arterioscler Thromb Vasc Biol 2006;26(7):1667–73.
[52] Ameziane N, et al. Association of the Toll-like receptor 4 gene Asp299Gly polymorphism with acute coronary events. Arterioscler Thromb Vasc Biol 2003;23:61–4.
[53] Horne BD, et al. Multiple-polymorphism associations of 7 matrix metalloproteinase and tissue inhibitor metalloproteinase genes with myocardial infarction and angiographic coronary artery disease. Am Heart J 2007;154(4):751–8.
[54] Wang Q, et al. Premature myocardial infarction novel susceptibility locus on chromosome 1P34-36 identified by genomewide linkage analysis. Am J Hum Genet 2004;74:262–71.
[55] The BHF Family Heart Study Research Group. A genome linkage study of 1933 families affected by premature coronary artery disease: the British Heart Foundation (BHF) Family Heart Study. Am J Hum Genet 2005;77:1011–20.
[56] Harrap SB, et al. Genome-wide linkage analysis of the acute coronary syndrome suggests a locus on chromosome 2. Arterioscler Thromb Vasc Biol 2002;22:874–8.
[57] Pajukanta P, et al. Two loci on chromosomes 2 and X for premature coronary heart disease identified in early- and late-settlement populations of Finland. Am J Hum Genet 2000;67:1481–93.
[58] Hauser ER, et al. A genomewide scan for early-onset coronary artery disease in 438 families: the GENECARD Study. Am J Hum Genet 2004;75:436–47.
[59] Helgadottir A, et al. The gene encoding 5-lipoxygenase activating protein confers risk of myocardial infarction and stroke. Nat Genet 2004;36:233–9.
[60] Broeckel U, et al. A comprehensive linkage analysis for myocardial infarction and its related risk factors. Nat Genet 2002;30:210–4.
[61] Wang L, et al. Mutation of MEF2A in an inherited disorder with features of coronary artery disease. Science 2003;302:1578–81.
[62] Francke S, et al. A genome-wide scan for coronary heart disease suggests in Indo-Mauritians a susceptibility locus on chromosome 16p13 and replicates linkage with the metabolic syndrome on 3q27. Hum Mol Genet 2001;10:2751–65.

[63] Farrall M, et al. Genomewide mapping of susceptibility to coronary artery disease identifies a novel replicated locus on chromosome 17. PLoS Genet 2006;2:e72.
[64] Chiodini BD, et al. Meta-analysis of 4 coronary heart disease genome-wide linkage studies confirms a susceptibility locus on chromosome 3q. Arterioscler Thromb Vasc Biol 2003;23:1863–8.
[65] Helgadottir A, et al. A variant of the gene encoding leukotriene A4 hydrolase confers ethnicity-specific risk of myocardial infarction. Nat Genet 2005;38:68–74.
[66] Assimes TL, et al. Common polymorphisms of *ALOX5* and *ALOX5AP* and risk of coronary artery disease. Hum Genet 2008;123:399–408.
[67] Lieb W, et al. Lack of association between the *MEF2A* gene and myocardial infarction. Circulation 2008;117:185–91.
[68] Shiffman D, et al. Gene variants of VAMP8 and HNRPUL1 are associated with early-onset myocardial infarction. Arterioscler Thromb Vasc Biol 2006;26:1613–8.
[69] Wang L, et al. Peakwide mapping on chromosome 3q13 identifies the kalirin gene as a novel candidate gene for coronary artery disease. Am J Hum Genet 2007;80:650–63.
[70] Teare ND, et al. Genetic linkage studies. Lancet 2008;366:1036–44.
[71] Gibbs RA, et al. The international HapMap project. Nature 2003;426:789–96.
[72] Altshuler D, et al. A haplotype map of the human genome. Nature 2005;2005(437):1299–320.
[73] McPherson R, et al. A common allele on chromosome 9 associated with coronary heart disease. Science 2007;316:1488–91.
[74] Helgadottir A, et al. A common variant on chromosome 9p21 affects the risk of myocardial infarction. Science 2007;316:1491–3.
[75] Wellcome Trust Case Control Consortium. Genome-wide association study of 14,000 cases of seven common diseases and 3,000 shared controls. Nature 2007;447:661–78.
[76] Samani NJ, et al. Genomewide association analysis of coronary artery disease. N Engl J Med 2007;2357:443–53.
[77] Schunkert H, et al. Repeated replication and a prospective meta-analysis of the association between chromosome 9p21.3 and coronary artery disease. Circulation 2008;117:1675–84.
[78] Assimes T, et al. Susceptibility locus for clinical and subclinical coronary artery disease at chromosome 9p21 in the multi-ethnic ADVANCE Study. Hum Genet 2008. [Epub ahead of print].
[79] Hinohara K, et al. Replication of the association between a chromosome 9p21 polymorphism and coronary artery disease in Japanese and Korean populations. J Hum Genet 2008;53: 357–9.
[80] Shen GQ, et al. Four SNPs on chromosome 9p21 in a South Korean population implicate a genetic locus that confers high cross-race risk for development of Coronary Artery Disease. Arterioscler Thromb Vasc Biol 2007;28:360–5.
[81] Zeggini E, et al. Replication of genome-wide association signals in Uk samples reveals risk loci for type 2 diabetes. Science 2007;316:1336–41.
[82] Scott LJ, et al. A genome-wide association study of type 2 diabetes in Finns detects multiple susceptibility variants. Science 2007;316:1341–5.
[83] Diabetes Genetics Initiative of Broad Institute of Harvard and MIT, Lund University, and Novartis Institutes of Biomedical Research, et al. Genome-wide association analysis identifies loci for type 2 diabetes and triglyceride levels. Science 2008;316:1331–6.
[84] Salonen JT, et al. Type 2 diabetes whole genome association study in four populations: the DiaGen consortium. Am J Hum Genet 2007;81:338–45.
[85] Sladek R, et al. A genome-wide association study identifies novel risk loci for type 2 diabetes. Nature 2007;445:881–5.

[86] Broadbent HM, et al. Susceptibility to coronary artery disease and diabetes is encoded by distinct, tightly linked, SNPs in the ANRIL locus on chromosome 9. Hum Mol Genet 2008;17:806–14.
[87] Pasmant E, et al. Characterisation of germ-line deletion, including the entire INK4/ARF locus in a melanoma-neural system tumor family: identification of ANRIL, an antisense non-coding RNA whose expression co-clusters with ARF. Cancer Res 2007;67:3963–9.
[88] Helgadottir A, et al. The same sequence variant on 9p21 associates with myocardial infarction, abdominal aortic aneurysm and intracranial aneurysm. Nat Genet 2008;40:217–24.
[89] Samani NJ, et al. Coronary artery disease-associated locus on chromosome 9p21 and early markers of atherosclerosis. Arterioscler Thromb Vasc Biol 2008. [Epub ahead of print].
[90] Lawlor DA. Mendelian randomization: using genes as instruments for making causal inferences in epidemiology. Stat Med 2008;2008(27):1133–63.
[91] Kivimaki M. Mendelian randomization suggests no causal association between C-reactive protein and carotid intima-media thickness in the young Finns study. Arterioscler Thromb Vasc Biol 2007;27:978–9.
[92] Davey Smith G. Does elevated plasma fibrinogen increase the risk of coronary heart disease? Evidence from a meta-analysis of genetic association studies. Arterioscler Thromb Vasc Biol 2005;25:2228–33.
[93] Ference B. Effect of long-term exposure to lower low-density lipoprotein cholesterol beginning early in life on the risk of coronary heart disease: a Mendelian randomization analysis. J Am Coll Cardiol 2012;60:2631–9.

Chapter 8

Role of Conventional Risk Factors in Genetic Susceptibility to Cardiovascular Diseases

D.I. Swerdlow[1,2], M. Michail[3], K. Zacharias[4]
[1]University College London, London, United Kingdom; [2]Imperial College London, London, United Kingdom; [3]Barts Health NHS Foundation Trust, London, United Kingdom; [4]Croydon University Hospital, London, United Kingdom

THE RELATIONSHIP OF RISK FACTORS AND GENETICS TO DISEASE

The large number of risk factors associated with cardiovascular disease (CVD) continues to increase and encompasses a broadening range of areas of biology, behavior, and environment. The relationships between risk factors can, however, take different forms (Fig. 8.1). First, a risk factor may directly cause the disease in question (Fig. 8.1A). Evidence for causation is difficult to demonstrate conclusively, as we discuss later, although it is widely accepted that such evidence is best derived from randomized controlled trials. An example of a risk factor with a strong causal effect is low-density lipoprotein cholesterol (LDL-C) in coronary artery disease (CAD). Second, levels of a risk factor may be influenced by the presence of the disease itself, in a phenomenon known as "reverse causation" (Fig. 8.1B). In such circumstances, the supposed risk factor cannot be said to have a causal role, but merely marks the presence of disease. An example of such a relationship is the association of C-reactive protein (CRP) with CAD risk, where CRP reliably associates with disease risk but does not play a causal role. Finally, both the risk factor and the disease endpoint may be related to a third factor, termed a "confounder," that is the true causal mediator of disease (Fig. 8.1C). The difficult task of determining causality in the presence of confounding factors will be further discussed later in this chapter.

Each of the types of relationship described before has important implications for the interpretation of an observed association between a risk factor and disease. Risk factors can act as predictors of future disease in otherwise healthy people. Risk estimation tools, such as the Framingham Risk Score or QRISK [1] use risk factors in this way to quantify an individual's likelihood of developing

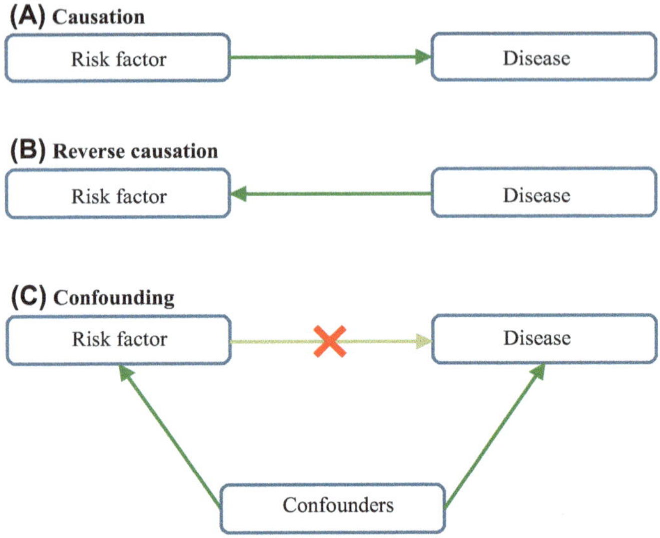

FIGURE 8.1 Relationships of risk factors with disease. A risk factor that has an observed association with disease risk may directly contribute to the causation of the disease endpoint (A). Alternatively, differences in the risk factor may merely reflect the presence of disease since the risk factor is influenced by the disease process itself, leading to a phenomenon known as "reverse causation" (B). Finally, a third variable related both to the risk factor and the disease may be the true causal mediator. This third variable is termed a confounder, and failure to recognize its presence may result in a causal role being erroneously ascribed to the initial risk factor (C).

disease within a given time period, such as the 5 or 10 years following the point of prediction. For the purposes of disease prediction, using such risk calculators, a risk factor need not play a causal role but must reliably associate with differences in disease risk.

An observed association of a risk factor with disease risk may suggest etiological mechanisms underlying disease, and in this case a causal relationship is necessary to infer a role for the risk factor in disease pathogenesis. Closely related to this is the value of risk factors for indicating biological pathways that might be modulated either pharmacologically or through other interventions to prevent and treat disease. In order to act as a therapeutic or preventive target a risk factor must play a causal role in disease development.

METHODS FOR ASSESSING CAUSALITY

A large proportion of risk factors identified by observational studies merely mark the presence of disease, rather than contributing to its development, and robust demonstration of causality to distinguish markers of disease from its etiological factors is one of the greatest challenges of epidemiology [2].

TABLE 8.1 Evidence for Relationships Between CVD Risk Factors and Disease Risk From Observational, Interventional, and Genetic Studies

	Association of Higher/Greater Risk Factor With CVD Risk		
Risk Factor	**Observational Studies**	**RCTs**	**Mendelian Randomization**
LDL-C	↑	↑	↑
HDL-C	↓	?	×
Triglycerides	↑	–	↑
CRP	↓	–	RP
Interleukin-6	↑	–	↑
Fibrinogen	↓	–	ib
BMI	↑	?	↑
Type 2 diabetes	↓	?	↓
Plasma glucose	↑	?	↑
Blood pressure	↓	↓	–
Alcohol consumption	?	–	↑
Tobacco smoking	↓	–	–

Key to symbols: ↑ Increase in level/risk; ↓ decrease in level/risk; ? Findings inconclusive; – Conclusive data unavailable; × No association.
BMI, Body mass index; *CRP*, C-reactive protein; *HDL-C*, high-density lipoprotein cholesterol; *LDL-C*, low-density lipoprotein cholesterol.

Broadly speaking, three types of study design in humans can be used to assess possible causal associations between a risk factor or disease marker and the development of disease. These are:

1. randomized controlled trials (RCTs)
2. observational studies
3. genetic studies or Mendelian randomization studies

The role of each of these methods in assessing established cardiovascular risk factor is summarized in Table 8.1. We will now explore them in more detail.

RANDOMIZED CONTROLLED TRIALS

Randomized controlled trials (RCTs) are generally considered the most reliable method of evaluating causality and have been successfully used in this regard historically. This is due to the features in the design of RCTs that are thought to limit the effect of confounders [3]. The act of comparing an intervention, whether it be a drug, an operation, or a behavioral modification, with a control allows observed differences between the intervention and control groups to be deemed consequences of that intervention.

At the core of the RCT design is the random allocation of trial participants to intervention or control groups. Since allocation is due to chance alone and allocation to either group has equal likelihood, potential confounders of the intervention–outcome relationship should be equally distributed between the two groups. That is to say, provided a sufficiently large sample is used, the two groups should be matched at baseline for features such as the proportion of male and female participants, mean age body weight, and so on. With the effects of possible confounding factors limited in this way, any differences in outcomes observed between the two groups can be ascribed to the intervention.

A second inherent feature of the RCT design is the prospective follow-up of the participants. This improves the chances of establishing a temporal sequence between exposure and outcome, which is one of the stipulations for causal inference proposed by Austin Bradford Hill [3]. Reverse causation is therefore avoided.

Finally, many RCTs employ blinding as a means of reducing a number of forms of bias, including measurement bias, observer bias, and recall bias. The trial investigators or participants or both may be kept unaware of ("blinded" to) the trial arm to which participants have been allocated. Although blinding is not always feasible, such as when the intervention is an invasive procedure, it is almost universal in modern drug trials, as it is a valuable means of increasing the internal validity of the trial findings.

The special role of the RCT in causal inference has led to the demonstration of efficacy of many treatments at the heart of CVD management. For instance, trials of statin therapy have confirmed a causal role for LDL-C in cardiovascular disease [4] and the drugs have had a substantial impact on population CVD risk. For many risk factors, however, trials have not yet been conducted, often because no safe pharmacological agent exists for use in humans, or because the utility of targeting the risk factor in question has not been shown. As we will discuss later, genetic studies can contribute valuable information in these areas.

OBSERVATIONAL STUDIES

Although observational studies are more prone to confounding and bias, they are a very useful tool in epidemiology as there may be important practical or ethical issues that preclude the design of an RCT in certain questions. For example, smoking is considered to be one of the most well-established risk factor for CVD, mainly as a result of large observational studies [5,6]. Allocation of patients to an intervention involving cigarette smoking would, however, clearly be unethical. In general, use of large, appropriately sampled data in observational studies allows for statistical adjustment for most measured confounding factors. However, residual measured and nonmeasured confounding may persist, biases may influence results, and reverse causation may distort true causal relationships. Inferring causality purely on the basis of observational studies is therefore challenging and often impossible.

GENETIC STUDIES AND MENDELIAN RANDOMIZATION

Searching the Genome for Risk-Associated Variants

Since the proliferation of genome-wide association (GWA) studies, genetic methods have begun to emerge as alternative means of conducting robust causal inference analysis. GWA studies continue to provide valuable insights into the genetic architecture underlying the pathogenesis of complex diseases such as CVD, and many important disease-related genetic loci have been reported in well-replicated studies (Table 8.1) [7]. Furthermore, much success has been made in identifying genetic variants associated with risk factors such as circulating lipids [8], blood pressure [9], and body mass index (BMI) [10], the biological determinants of which had previously been poorly or incompletely understood. Both disease- and risk factor–associated genetic variants can be used to investigate the causal relevance of risk factors to common, complex diseases such as CVD, and to investigate novel pathways that might be modulated to reduce disease risk or treat established disease. Here we will examine the contribution that genetic studies have made to exploration of the roles of established and emerging risk factors in CVD.

Identifying Causal Risk Factors

While GWA studies have generated a growing body of genetic variants associated with complex diseases and their related biomarkers, other methods have been required to exploit those findings for robust causal inference. Chief among these methods is Mendelian randomization (MR) [11,12]. MR analysis has been termed "nature's randomized trial" [13] and uses genetic variants associated with a putative risk factor as unconfounded proxies for that risk factor in large epidemiological studies in order to determine the causal relevance of the risk factor to a given disease. In the context of MR, markers of genetic variation are generally single nucleotide polymorphisms (SNPs) since these sequence variants occur commonly throughout the genome, are frequently found in protein-coding regions, and are highly conserved between generations. From a practical point of view, SNP genotypes are relatively quick, straightforward and cost-effective to measure, and detection of genotype at over 1,000,000 SNPs in an individual is currently possible for under $50, with many individuals being genotyped rapidly in parallel.

The MR model considers three relationships (Fig. 8.2) [11]. First, the observed association between the risk factor and risk of disease must be

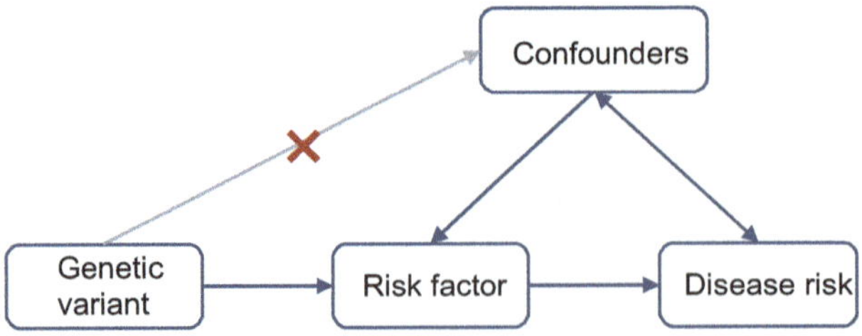

FIGURE 8.2 The Mendelian randomization model. Mendelian randomization examines three relationships in order to evaluate the causal contribution of a risk factor to a disease endpoint. First, the risk factor must show an association with disease risk in observational studies. Second, a genetic variant is selected that associates with levels of the risk factor, but not with potential confounders of the risk factor–disease relationship. The genetic variant therefore acts as an unconfounded, unbiased instrument for the risk factor. If the risk factor–disease, gene variant–risk factor, and gene variant–disease relationships can be shown to exist, the genetic variant is assumed to exert its effect on risk of disease exclusively through its action on risk factor. The risk factor is therefore ascribed a causal role in disease.

confirmed using classical epidemiological methods. Second, the genetic variant must be associated with the risk factor but not with potential confounders of the risk factor–disease relationship. Finally, the genetic instrument must be associated with risk of disease, accompanied by the assumption that this relationship acts only via the effect of the risk factor on disease risk. If all three of these relationships can be demonstrated with statistical reliability then a causal role can be ascribed to the risk factor in the disease (Fig. 8.2).

The MR paradigm rests on some important features of genetic variation, derived from those originally described by Gregor Mendel after his fundamental botanical experiments [14]. Key to these is that the genotype of an individual at a given locus in the genome is allocated at random from parental gametes at the time of conception. Genotype can therefore be considered analogous to treatment allocation in an RCT. Further, genotype at one locus is generally inherited independently of genotype at other loci, and genotype remains unchanged once it is determined at the time of conception. Therefore, as in the RCT, differences in phenotype observed between genotype groups at one locus (eg, GG carriers vs AA carriers) can be attributed to the biological consequences of variation in that gene. These consequences may be differences in the quantity or function of the protein product of the gene, the action of a transcription factor, or another feature of transcriptional or translational regulation. The similarities between MR and the classical RCT design are strong—both involve an "intervention" that is randomly allocated in order to define active and control groups between which phenotypic differences can be observed and quantified. Although the structures of MR and RCT analyses are similar, it is should be noted that there are important points of difference. The magnitudes of associations between common genetic variants and risk factors or disease are generally substantially

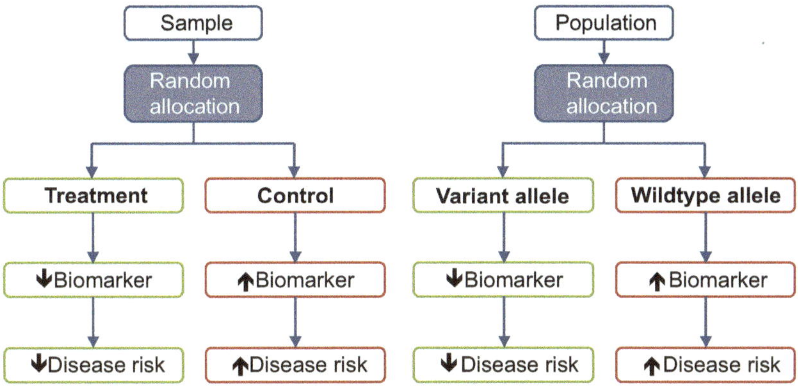

FIGURE 8.3 Natural parallels between the randomized controlled trial and Mendelian randomization. The Mendelian randomization (MR) paradigm shares a number of features with the randomized controlled trial that allow both types of study to be used for causal inference. Both models involve an intervention. In the RCT this may be a drug or an invasive procedure; in the MR study the genotype is the naturally occurring intervention. Differences observed between trial treatment groups or between genotype groups can therefore be deemed consequences of the intervention. In each case, individuals are randomly allocated to an intervention—in the RCT through a conventional randomization process and in the MR as a consequence of the laws of inheritance. Randomization in both cases enables balancing of potential confounders between the study groups. RCT and MR analyses are necessarily longitudinal in design, with a clear temporal sequence from intervention to its observed consequences. As a result, reverse causation is largely avoided. Finally, in many RCTs and in most MR analyses, participants and investigators are unaware of the study group to which they have been allocated, therefore reducing the influence of bias.

smaller (often by two or three orders of magnitude) than those reported in RCTs of highly potent pharmacological mediators of biological pathways. Consequently, very large samples are often required for MR analysis in order to achieve sufficient statistical power, with studies frequently including data for over 150,000 individuals. Although not the focus of this discussion, an extensive body of literature is growing on methodological considerations in MR [15,16] (Fig. 8.3).

In the rest of this chapter, we will discuss how genetic studies have provided information about the roles of key risk factors in CVD and how such genetic studies have helped to address controversies in this area.

BLOOD LIPIDS

LDL Cholesterol

Circulating LDL-C provides a helpful test case for the role of genetic studies in investigation of CVD risk factors. Observational data show very clearly the positive relationship of circulating LDL-C with CVD risk [17]. Furthermore, a wealth of data from RCTs of lipid-lowering statin therapy across a spectrum of baseline cardiovascular risk and concomitant risk factors has demonstrated through interventional studies a causal role for LDL-C in CVD risk. Blood lipids are among the phenotypes most intensively studied by GWA studies, and large numbers of loci associated with circulating concentrations of total, LDL- and high-density lipoprotein (HDL) cholesterol (HDL-C) have been reported [8]. These variants have been used in several MR analyses to confirm the causal role of higher circulating concentrations of LDL-C in CVD [18,19]. LDL-C is privileged among circulating lipids in that it has an extensively studied and well-characterized pharmacological mediator in the shape of the HMG-CoA reductase inhibitors, or statins. The majority of large RCTs of statin therapy preceded MR studies of LDL-C in coronary heart disease (CHD), and so direct comparison of the directions of effect observed in RCTs and MR studies has been possible. Reassuringly, genetic studies, often using composite scores incorporating information on a number of LDL-C-associated genetic variants, have confirmed the causal role of higher circulating LDL-C concentrations in CHD risk that was demonstrated by the statin trials. Closely related to LDL-C is lipoprotein(a) (Lp(a)), which is also positively associated with risk of CHD in observational studies [20]. Unlike LDL-C, however, there is no specific pharmacological mediator of Lp(a) in clinical use. Apolipoprotein(a), which is found on the Lp(a) particle, is the direct protein product of a single gene locus (*LPA*, chromosome 6p26) and is a reliable marker of Lp(a) concentration [21]. In contrast, circulating concentrations of the complex LDL-C molecule are influenced by at least 56 gene loci [8]. MR analyses investigating Lp(a) therefore used single variants at the *LPA* locus to demonstrate a causal role for Lp(a) in both CHD [22] and aortic valve disease [23].

HDL Cholesterol and Triglycerides

Greater controversy surrounds the roles of other lipid fractions in CVD. Plasma triglyceride concentrations exhibit a clear positive relationship with CVD risk in observational studies [17] but no specific triglyceride-lowering drug treatment is widely available. An early MR study used a variant at the *APOA5* locus that associated strongly with triglyceride concentrations and concluded that pathways involving triglycerides and apolipoprotein A-V played a causal role in CHD [24]. A later, more intricate analysis using a composite score of triglyceride-associated variants [25], and other studies using variants in *APOC3* [26,27] reinforced a likely causal role for triglycerides in CHD. Far greater uncertainty is found around the role of HDL-C. Once again, observational evidence is strong for the relationship between higher HDL-C and lower CVD risk [17], and HDL-C has long been proposed as a protective factor in CVD, with a causal role in disease suggested but not yet proven. Several attempts have been made to investigate the causal role of HDL-C in CVD using MR. These have included analyses using single SNPs in genes involved in HDL-C metabolism such as *LIPG* [28], *LCAT* [29], and *ABCA1* [30], and others that employ allele scores incorporating information from several loci [18,28]. Interestingly, despite the consistency of observational data, none of these MR studies has offered strong evidence for a causal role of HDL-C in CVD. A great deal of debate has stemmed from these findings, particularly regarding the possibility that different functions or subforms of HDL-C particles may contribute in a causal fashion to CVD risk [28]. Adding to clinical relevance of the HDL-C debate is the ongoing development of drugs that inhibit cholesteryl ester transfer protein (CETP), a key mediator of HDL-C metabolism. Of these drugs, one (torcetrapib) was abandoned because of adverse effects on blood pressure in a large RCT leading to a higher CVD death rate among torcetrapib-treated individuals [31]; and development of another (dalcetrapib) was terminated after a large RCT because of lack of efficacy [32]. Two other agents (anacetrapib [33] and evacetrapib [34]) remain in development and hope remains that they may provide clinical benefit in CHD prevention [35]. The nature of the role of HDL-C in CVD remains uncertain, but to date MR studies addressing the question from a number of biological viewpoints have yet to provide strong evidence in favor of a causal role [36].

INFLAMMATION

Circulating markers and mediators of inflammation have emerged as intriguing risk factors for CVD. As with lipids, interest in inflammation has been reinforced by observational epidemiology, which has proposed positive associations of biomarkers such as C-reactive protein (CRP) [37], interleukin-6 (IL-6) [38], and fibrinogen [39] with CVD risk. Given the biological plausibility supporting inflammation as an etiological mechanism in atherogenesis [40], substantial interest has grown in inflammatory pathways as targets for prevention and treatment of CVD. Each of the three aforementioned biomarkers of inflammation has been investigated using MR, with some interesting findings. CRP had perhaps the highest profile since the CRP-lowering effects of statin therapy had previously been proposed as synergistic with the lipid-lowering effects of the drugs in CHD risk reduction [41]. Unfortunately, no specific pharmacological inhibitor of CRP is available for evaluation in an RCT, leaving an opportunity for MR to offer information about the causal role of CRP. A number of MR analyses have addressed this issue using common variants in the gene encoding CRP (*CRP*, chromosome 1q23). In each such MR study, of which the largest included over 46,000 CHD cases [42], the genetic variants have associated reliably with circulating CRP concentrations but have shown no associations with CHD risk [42–44], providing strong evidence against a causal role for CRP in CHD. Similarly, an MR study of fibrinogen using a variant in its encoding gene (*FGB*, chromosome 4q28) reported strong associations with circulating fibrinogen concentrations but none with CHD risk [45]. Findings for IL-6 signaling have been more promising. IL-6 is a proinflammatory cytokine with a number of biological functions of relevance to atherogenesis and ischemic cardiovascular events [46]. Commonly occurring variants in the gene encoding the IL-6 receptor (*IL6R*, chromosome 1q21) are reliable determinants of circulating IL-6 concentration and include a functional variant that influences the signaling capability of the receptor. These variants have been used in large MR studies to show that attenuated IL-6 signaling via its receptor has a causally protective role in CHD [47,48], suggesting that the pathway may be a therapeutic or preventive drug target.

ANTHROPOMETRIC TRAITS AS RISK FACTORS

MR studies have addressed the causal relevance of classical risk factors in CVD beyond circulating biomarkers. Greater adiposity or BMI has long been proposed as an important cardiovascular risk factor, and again we see strong observational evidence in this regard [49]. Randomized trials of weight reduction have been relatively few in number and challenging to conduct leaving the question of causality unanswered. BMI is a complex trait with many contributory genetic loci, a large number of which have been reported by GWA studies [10,50]. MR studies have employed multiple variant allele scores, similar in principle to those used in the MR studies of lipids described previously. Each of these MR studies has supported a causal role for higher BMI in raising risk of CHD events [51–53], reinforcing widespread public health efforts to promote weight loss in the population. Closely related to BMI is type 2 diabetes (T2DM), which is generally accepted to be a potent risk factor for CVD [54]. GWA studies have again identified a wide range of genetic loci associated with risk of T2DM [55], and with its related biomarkers such as plasma glucose, insulin, glycosylated hemoglobin (HbA_{1c}), and markers of insulin resistance (eg, HOMA-IR) [56]. These variants have been combined into genetic risk scores for plasma glucose, HbA_{1c}, and diabetes risk itself in a number of MR studies, each of which supports a causal role for dysglycemia in increasing risk of CVD [57–59]. Controversy persists, however, on the protective role for CVD of oral hypoglycemic agents in patients with T2DM [60].

Blood pressure is unusual among conventional cardiovascular risk factors since it has been extensively investigated by GWA studies [9] yet has featured little in the CVD MR literature, perhaps because data from RCTs of antihypertensive therapy have reliably confirmed the causal role of higher blood pressure in CVD [61] (Table 8.1). That said, several MR studies have addressed the roles of other risk factors in the development of hypertension. For example, higher BMI has a causal effect on higher blood pressure [53], and higher circulating uric acid leads to lower blood pressure, with a potential interaction with diuretic therapy [62]. There may be scope for use of MR methods to investigate further the potential effects of blood pressure modification in other disease areas.

BEHAVIORAL AND ENVIRONMENTAL RISK FACTORS

The risk factors explored previously can be termed endogenous risk factors, that is to say they originate and operate within the physiological system of the human body. There are many important exogenous risk factors, including behavioral and environmental factors. Among these, two of the most important are alcohol consumption and tobacco smoking.

Alcohol

The relationship of alcohol consumption to CVD risk has been the subject of debate for many years, chiefly on account of the apparently nonlinear association with risk reported by observational epidemiological studies. Individuals who drink moderate amounts of alcohol appear to have lower risk of CVD than those who drink none at all, while heavier drinkers seem to have the highest risk, resulting in the so-called J- or U-shaped curve of association [63]. Theoretically, the most robust means of determining whether the alcohol-CVD relationship is indeed nonlinear, and whether moderate alcohol consumption conveys some protective benefit in CVD is through an RCT, but the design of such a study poses considerable ethical problems. Genetic studies using the Mendelian randomization model offer an opportunity to address the question with a similar level of reliability to an intervention trial. Alcohol, in the form of ethanol, is metabolized by hepatic dehydrogenase enzymes including alcohol dehydrogenase 1B (encoded by *ADH1B*, chromosome 4q23) and aldehyde dehydrogenase 2 (the product of *ALDH2*, chromosome 12q24). Carriage of functional variants in either gene, which are particularly common in people of East Asian ancestry, leads to an increase in the adverse effects of alcohol consumption, such as flushing and nausea, and consequently to lower alcohol consumption [64,65]. To frame this in the context of an RCT, random allocation to a particular genotype leads to greater or less tolerance of alcohol and therefore to demonstrably higher or lower alcohol consumption.

Under the terms of the Mendelian randomization principle, these genetic variants can be used as unconfounded instruments for alcohol consumption, leading to groups with different levels of alcohol consumption determined by randomly assigned genotype. This technique has been successfully employed in large population-based samples to demonstrate that carriers of alleles leading to higher alcohol consumption had higher risk of CVD events. Furthermore, rather than a U-shaped association, alcohol has a log-linear, causal relationship with CVD risk. That is to say, any increase in alcohol consumption confers higher risk of a CVD event [66,67]. The largest of these studies [66] interrogated the associations of the genetic variants with other cardiovascular risk factors and reported that the variant association with greater alcohol consumption was also associated with higher systolic blood pressure, circulating interleukin-6 concentration, and BMI, providing some insight into possible mechanisms through which alcohol may raise CVD risk. An important corollary of these findings for public health and health promotion is that even modest reduction in alcohol consumption is likely to reduce an individual's risk of cardiovascular events.

Smoking

Tobacco smoking has long been implicated as a risk factor in CVD, with very strong supporting evidence from observational studies. The causality of the relationship is assumed since, as with alcohol, a randomized trial of smoking is impossible. The pathological pathways underlying the smoking-CVD relationship have, however, come under scrutiny and here too genetics has helped to provide further information. GWA studies of smoking behaviors have reported association with variants at a number of loci [68,69]. Among these are polymorphisms at loci with links to biological plausible mechanisms such as *CHRNA5* (chromosome 15q24), encoding a nicotinic acetylcholine receptor and *CYP2A6* (chromosome 19q13) a hepatic enzyme involved in the metabolism of nicotine. An MR study investigating the relationships between smoking and cardiovascular risk factors reported that a variant in *CHRNA5* was associated with heavier smoking and with lower waist and hip circumferences and systolic blood pressure, and higher HDL-C concentrations and resting heart rate [70]. The association of heavier smoking with lower BMI was similarly reported by an independent population study using a variant in *CHRNA3* [71]. These findings contribute only modestly to our understanding of the pathological mechanisms linking smoking with CVD. They and the MR studies of alcohol consumption do, however, make a strong case for the use of genetic studies to investigate the causal relevance of behavioral and environmental risk factors to complex diseases such as CHD where RCTs are not feasible. Establishing causality in a risk factor–disease association lends validity to efforts to modify that risk factor and therefore has important implications for public and individual health improvement.

LIMITATIONS OF CAUSAL INFERENCE USING GENETICS

Herein, we have described some of the more important contributions of MR studies to understanding of CVD and its risk factors, but the limitations of the MR technique should not be overlooked [72]. For some risk factors, whether they be circulating biomarkers or behavioral traits, genetic variants associated with the trait may not be available. In such an instance, an appropriate GWA study may not have been conducted or may not have detected any associated variants. Even when variants associated with the trait of interest have been reliably identified, these may be unsuitable for use in MR analyses because they also associate with a number of other traits that may confound the index association. A structural phenomenon in the genome, known as linkage disequilibrium, results in loci in close physical proximity on a chromosome being inherited together. In such a case the phenotypic effects of one locus may be indistinguishable from a second in strong linkage disequilibrium, and trait associations of one gene may be confounded by those of the second. Linkage disequilibrium structures can differ markedly between populations of different ancestries, such that loci are related differently to one another between ancestral groups. Therefore, a gene–risk factor association observed in a sample of individuals of European ancestry may not be replicable in individuals of African descent because of differences in linkage disequilibrium. As a result, MR studies are generally conducted in samples from a single ancestral group. Many early GWA and MR studies used samples drawn from populations of European ancestry, with limited representation of other ancestral groups in the published literature. More recently studies are including a greater range of ancestral groups, providing greater insight both into risk factor–disease relationships and the specific nature of these relationships in different populations. As noted before, common genetic variants tend to have small effects on measured phenotypes [73]. For example, the variant allele of a polymorphism in the *HMGCR* gene, encoding HMG-CoA reductase (the target of statin drugs), is present in approximately 25% of people of European ancestry. The same variant is associated with a difference in circulating LDL-C of 0.08 mmol/L [74]. When compared with statin therapy that can reduce LDL-C by up to 2.00 mmol/L [75], the modest size of common genetic effects is clear. The samples required for genetic studies such as Mendelian randomization are therefore very large in order to achieve appropriate statistical power, and analyses frequently include data for over 200,000 individuals [66,74]. Perhaps the most important consideration in MR studies is the possibility that the phenotypic effect of a genetic variant, although present throughout the life course, may be modulated by the effects of other genes or behaviors as an individual ages. This applies particularly to deleterious variants, whose effects may be attenuated to reduce harm to the individual. Consequently, phenotypic effects of variants observed later in life may be different from those in adolescence and early adulthood.

CONCLUSIONS

We have discussed the contribution of genetic studies to the understanding of roles of a wide range of risk factors in coronary disease. From confirmatory analysis in proatherogenic lipids and building a case against C-reactive protein as a causal mediator in CHD, to investigation of complex behavioral phenotypes, genetic studies have become a powerful tool for examining CVD risk. The Mendelian randomization principle has been extensively developed and refined such that, notwithstanding its limitation, it offers opportunities for causal inference that were previously impossible. The scope for further genetic research into the causal contributions of the vast range of CVD risk factors is great, and judiciously designed studies are proving an efficient and fruitful means of addressing these important questions.

REFERENCES

[1] Hippisley-Cox J, Coupland C, Vinogradova Y, Robson J, Minhas R, Sheikh A, et al. Predicting cardiovascular risk in England and Wales: prospective derivation and validation of QRISK2. BMJ June 28, 2008;336(7659):1475–82.

[2] Smith GD, Lawlor DA, Harbord R, Timpson N, Day I, Ebrahim S. Clustered environments and randomized genes: a fundamental distinction between conventional and genetic epidemiology. PLoS Med December 2007;4(12):e352.

[3] Hill AB. The environment and disease: association or causation? Proc R Soc Med May 1965;58:295–300.

[4] Baigent C, Blackwell L, Emberson J, Holland LE, Reith C, Bhala N, et al. Efficacy and safety of more intensive lowering of LDL cholesterol: a meta-analysis of data from 170,000 participants in 26 randomised trials. Lancet November 13, 2010;376(9753):1670–81.

[5] Doll R, Peto R. Mortality in relation to smoking: 20 years' observations on male British doctors. Br Med J December 25, 1976;2(6051):1525–36.

[6] Doll R, Gray R, Hafner B, Peto R. Mortality in relation to smoking: 22 years' observations on female British doctors. Br Med J April 5, 1980;280(6219):967–71.

[7] Swerdlow DI, Hingorani AD, Humphries SE. Genetic risk factors and mendelian randomization in cardiovascular disease. [Internet] Curr Cardiol Rep May 2015;17(5). [cited 24.06.15.]; Available from: http://dx.doi.org/10.1007/s11886-015-0584-x.

[8] Willer CJ, Schmidt EM, Sengupta S, Peloso GM, Gustafsson S, Kanoni S, et al. Discovery and refinement of loci associated with lipid levels. Nat Genet October 6, 2013;45(11):1274–83.

[9] Wain LV, Verwoert GC, O'Reilly PF, Shi G, Johnson T, Johnson AD, et al. Genome-wide association study identifies six new loci influencing pulse pressure and mean arterial pressure. Nat Genet October 2011;43(10):1005–11.

[10] Speliotes EK, Willer CJ, Berndt SI, Monda KL, Thorleifsson G, Jackson AU, et al. Association analyses of 249,796 individuals reveal 18 new loci associated with body mass index. Nat Genet November 2010;42(11):937–48.

[11] Lawlor DA, Harbord RM, Sterne JAC, Timpson N, Davey Smith G. Mendelian randomization: using genes as instruments for making causal inferences in epidemiology. Stat Med April 15, 2008;27(8):1133–63.

[12] Ebrahim S, Davey Smith G. Mendelian randomization: can genetic epidemiology help redress the failures of observational epidemiology? Hum Genet February 2008;123(1):15–33.

[13] Hingorani A, Humphries S. Nature's randomised trials. Lancet December 3, 2005;366 (9501):1906–8.

[14] Mendel G. Experiments in plant-hybridisation. Brunn Nat Hist Soc 1865.

[15] Burgess S, Timpson NJ, Ebrahim S, Davey Smith G. Mendelian randomization: where are we now and where are we going? Int J Epidemiol April 2015;44(2):379–88.

[16] Burgess S, Scott RA, Timpson NJ, Davey Smith G, Thompson SG. EPIC-InterAct Consortium. Using published data in Mendelian randomization: a blueprint for efficient identification of causal risk factors. Eur J Epidemiol March 15, 2015.

[17] Di Angelantonio E, Sarwar N, Perry P, Kaptoge S, Ray KK, Thompson A, et al. Major lipids, apolipoproteins, and risk of vascular disease. JAMA J Am Med Assoc November 11, 2009;302(18):1993–2000.

[18] Holmes MV, Asselbergs FW, Palmer TM, Drenos F, Lanktree MB, Nelson CP, et al. Mendelian randomization of blood lipids for coronary heart disease. Eur Heart J January 27, 2014.

[19] Teslovich TM, Musunuru K, Smith AV, Edmondson AC, Stylianou IM, Koseki M, et al. Biological, clinical and population relevance of 95 loci for blood lipids. Nature August 5, 2010;466(7307):707–13.

[20] Erqou S, Kaptoge S, Perry PL, Di Angelantonio E, Thompson A, White IR, et al. Lipoprotein(a) concentration and the risk of coronary heart disease, stroke, and nonvascular mortality. JAMA J Am Med Assoc July 22, 2009;302(4):412–23.

[21] Ober C, Nord AS, Thompson EE, Pan L, Tan Z, Cusanovich D, et al. Genome-wide association study of plasma lipoprotein(a) levels identifies multiple genes on chromosome 6q. J Lipid Res May 2009;50(5):798–806.

[22] Clarke R, Peden JF, Hopewell JC, Kyriakou T, Goel A, Heath SC, et al. Genetic variants associated with Lp(a) lipoprotein level and coronary disease. N Engl J Med December 24, 2009;361(26):2518–28.

[23] Kamstrup PR, Tybjærg-Hansen A, Nordestgaard BG. Elevated lipoprotein(a) and risk of aortic valve stenosis in the general population. J Am Coll Cardiol February 11, 2014;63(5):470–7.

[24] Sarwar N, Sandhu MS, Ricketts SL, Butterworth AS, Di Angelantonio E, Boekholdt SM, et al. Triglyceride-mediated pathways and coronary disease: collaborative analysis of 101 studies. Lancet May 8, 2010;375(9726):1634–9.

[25] Do R, Willer CJ, Schmidt EM, Sengupta S, Gao C, Peloso GM, et al. Common variants associated with plasma triglycerides and risk for coronary artery disease. Nat Genet November 2013;45(11):1345–52.

[26] Jørgensen AB, Frikke-Schmidt R, Nordestgaard BG, Tybjærg-Hansen A. Loss-of-function mutations in APOC3 and risk of ischemic vascular disease. N Engl J Med July 3, 2014;371(1):32–41.

[27] TG and HDL Working Group of the Exome Sequencing Project, National Heart, Lung, and Blood Institute, Crosby J, Peloso GM, Auer PL, Crosslin DR, et al. Loss-of-function mutations in APOC3, triglycerides, and coronary disease. N Engl J Med July 3, 2014;371(1): 22–31.

[28] Voight BF, Peloso GM, Orho-Melander M, Frikke-Schmidt R, Barbalic M, Jensen MK, et al. Plasma HDL cholesterol and risk of myocardial infarction: a mendelian randomisation study. [Internet] Lancet May 17, 2012. [cited 02.06.12]; Available from: http://www.ncbi.nlm.nih.gov/pubmed/22607825.

[29] Haase CL, Tybjærg-Hansen A, Qayyum AA, Schou J, Nordestgaard BG, Frikke-Schmidt R. LCAT, HDL cholesterol and ischemic cardiovascular disease: a Mendelian randomization study of HDL cholesterol in 54,500 individuals. J Clin Endocrinol Metab February 2012;97(2):E248–56.

[30] Frikke-Schmidt R. Genetic variation in the ABCA1 gene, HDL cholesterol, and risk of ischemic heart disease in the general population. Atherosclerosis February 2010;208(2):305–16.

[31] Barter PJ, Caulfield M, Eriksson M, Grundy SM, Kastelein JJP, Komajda M, et al. Effects of torcetrapib in patients at high risk for coronary events. N Engl J Med November 22, 2007;357(21):2109–22.

[32] Schwartz GG, Olsson AG, Abt M, Ballantyne CM, Barter PJ, Brumm J, et al. Effects of dalcetrapib in patients with a recent Acute coronary Syndrome. N Engl J Med November 29, 2012;367(22):2089–99.

[33] Cannon CP, Shah S, Dansky HM, Davidson M, Brinton EA, Gotto AM, et al. Safety of anacetrapib in patients with or at high risk for coronary heart disease. N Engl J Med December 16, 2010;363(25):2406–15.

[34] Nicholls SJ, Brewer HB, Kastelein JJP, Krueger KA, Wang M-D, Shao M, et al. Effects of the CETP inhibitor evacetrapib administered as monotherapy or in combination with statins on HDL and LDL cholesterol: a randomized controlled trial. [Internet] JAMA November 16, 2011;306(19). [cited 24.06.15]; Available from: http://jama.jamanetwork.com/article.aspx?doi=10.1001/jama.2011.1649.

[35] Rader DJ, deGoma EM. Future of cholesteryl ester transfer protein inhibitors. Annu Rev Med 2014;65:385–403.

[36] Harrison SC, Holmes MV, Humphries SE. Mendelian randomisation, lipids, and cardiovascular disease. [Internet] Lancet May 16, 2012. [cited 02.06.12]; Available from: http://www.ncbi.nlm.nih.gov/pubmed/22607824.

[37] Kaptoge S, Di Angelantonio E, Lowe G, Pepys MB, Thompson SG, Collins R, et al. C-reactive protein concentration and risk of coronary heart disease, stroke, and mortality: an individual participant meta-analysis. Lancet January 9, 2010;375(9709):132–40.
[38] Danesh J, Kaptoge S, Mann AG, Sarwar N, Wood A, Angleman SB, et al. Long-term interleukin-6 levels and subsequent risk of coronary heart disease: two new prospective studies and a systematic review. PLoS Med April 8, 2008;5(4):e78.
[39] Danesh J, Collins R, Appleby P, Peto R. Association of fibrinogen, C-reactive protein, albumin, or leukocyte count with coronary heart disease: meta-analyses of prospective studies. JAMA J Am Med Assoc May 13, 1998;279(18):1477–82.
[40] Hansson GK. Inflammation, atherosclerosis, and coronary artery disease. N Engl J Med April 21, 2005;352(16):1685–95.
[41] Ridker PM, Danielson E, Fonseca FAH, Genest J, Gotto Jr AM, Kastelein JJP, et al. Rosuvastatin to prevent vascular events in men and women with elevated C-reactive protein. N Engl J Med November 20, 2008;359(21):2195–207.
[42] CRP CHD Genetics Collaboration. Collaborative pooled analysis of data on C-reactive protein gene variants and coronary disease: judging causality by Mendelian randomisation. Eur J Epidemiol 2008;23(8):531–40.
[43] Casas JP, Shah T, Cooper J, Hawe E, McMahon AD, Gaffney D, et al. Insight into the nature of the CRP-coronary event association using Mendelian randomization. Int J Epidemiol August 2006;35(4):922–31.
[44] Elliott P, Chambers JC, Zhang W, Clarke R, Hopewell JC, Peden JF, et al. Genetic Loci associated with C-reactive protein levels and risk of coronary heart disease. JAMA J Am Med Assoc July 1, 2009;302(1):37–48.
[45] Keavney B, Danesh J, Parish S, Palmer A, Clark S, Youngman L, et al. Fibrinogen and coronary heart disease: test of causality by "Mendelian randomization". Int J Epidemiol August 2006;35(4):935–43.
[46] Naka T, Nishimoto N, Kishimoto T. The paradigm of IL-6: from basic science to medicine. Arthritis Res 2002;4(Suppl. 3):S233–42.
[47] Swerdlow DI, Holmes MV, Kuchenbaecker KB, Engmann J, Shah T, Sofat R, et al. The interleukin-6 receptor as a potential target for coronary heart disease prevention: evaluation using Mendelian randomisation. Lancet 2012.
[48] Sarwar N, Butterworth AS, Freitag DF, Gregson J, Willeit P, Gorman DN, et al. Interleukin-6 receptor pathways in coronary heart disease: a collaborative meta-analysis of 82 studies. Lancet March 31, 2012;379(9822):1205–13.
[49] Prospective Studies Collaboration, Whitlock G, Lewington S, Sherliker P, Clarke R, Emberson J, et al. Body-mass index and cause-specific mortality in 900,000 adults: collaborative analyses of 57 prospective studies. Lancet March 28, 2009;373(9669):1083–96.
[50] Berndt SI, Gustafsson S, Mägi R, Ganna A, Wheeler E, Feitosa MF, et al. Genome-wide meta-analysis identifies 11 new loci for anthropometric traits and provides insights into genetic architecture. Nat Genet May 2013;45(5):501–12.
[51] Nordestgaard BG, Palmer TM, Benn M, Zacho J, Tybjærg-Hansen A, Davey Smith G, et al. The effect of elevated body mass index on ischemic heart disease risk: causal estimates from a mendelian randomisation approach. PLoS Med May 2012;9(5):e1001212.
[52] Holmes MV, Lange LA, Palmer T, Lanktree MB, North KE, Almoguera B, et al. Causal effects of body mass index on cardiometabolic traits and events: a mendelian randomization analysis. Am J Hum Genet January 23, 2014.
[53] Fall T, Hägg S, Mägi R, Ploner A, Fischer K, Horikoshi M, et al. The role of adiposity in cardiometabolic traits: a mendelian randomization analysis. PLoS Med 2013;10(6):e1001474.

[54] Emerging Risk Factors Collaboration, Seshasai SRK, Kaptoge S, Thompson A, Di Angelantonio E, Gao P, et al. Diabetes mellitus, fasting glucose, and risk of cause-specific death. N Engl J Med March 3, 2011;364(9):829–41.

[55] Morris AP, Voight BF, Teslovich TM, Ferreira T, Segrè AV, Steinthorsdottir V, et al. Large-scale association analysis provides insights into the genetic architecture and pathophysiology of type 2 diabetes. Nat Genet September 2012;44(9):981–90.

[56] Scott RA, Lagou V, Welch RP, Wheeler E, Montasser ME, Luan J, et al. Large-scale association analyses identify new loci influencing glycemic traits and provide insight into the underlying biological pathways. Nat Genet August 12, 2012;44(9):991–1005.

[57] Ahmad OS, Morris JA, Mujammami M, Forgetta V, Leong A, Li R, et al. A Mendelian randomization study of the effect of type-2 diabetes on coronary heart disease. Nat Commun 2015;6:7060.

[58] Ross S, Gerstein HC, Eikelboom J, Anand SS, Yusuf S, Paré G. Mendelian randomization analysis supports the causal role of dysglycaemia and diabetes in the risk of coronary artery disease. Eur Heart J June 14, 2015;36(23):1454–62.

[59] Benn M, Tybjaerg-Hansen A, McCarthy MI, Jensen GB, Grande P, Nordestgaard BG. Nonfasting glucose, ischemic heart disease, and myocardial infarction: a Mendelian randomization study. J Am Coll Cardiol June 19, 2012;59(25):2356–65.

[60] Ferrannini E, DeFronzo RA. Impact of glucose-lowering drugs on cardiovascular disease in type 2 diabetes. Eur Heart J June 10, 2015.

[61] Blood Pressure Lowering Treatment Trialists' Collaboration, Ninomiya T, Perkovic V, Turnbull F, Neal B, Barzi F, et al. Blood pressure lowering and major cardiovascular events in people with and without chronic kidney disease: meta-analysis of randomised controlled trials. BMJ 2013;347:f5680.

[62] Sedaghat S, Pazoki R, Uitterlinden AG, Hofman A, Stricker BHC, Ikram MA, et al. Association of uric acid genetic risk score with blood pressure: the Rotterdam study. Hypertension November 2014;64(5):1061–6.

[63] Ronksley PE, Brien SE, Turner BJ, Mukamal KJ, Ghali WA. Association of alcohol consumption with selected cardiovascular disease outcomes: a systematic review and meta-analysis. BMJ 2011;342:d671.

[64] Macgregor S, Lind PA, Bucholz KK, Hansell NK, Madden PAF, Richter MM, et al. Associations of ADH and ALDH2 gene variation with self report alcohol reactions, consumption and dependence: an integrated analysis. Hum Mol Genet February 1, 2009;18(3):580–93.

[65] Yokoyama A, Yokoyama T, Mizukami T, Matsui T, Kimura M, Matsushita S, et al. Blood ethanol levels of nonabstinent Japanese alcoholic men in the morning after drinking and their ADH1B and ALDH2 genotypes. Alcohol Alcohol January 1, 2014;49(1):31–7.

[66] Holmes MV, Dale CE, Zuccolo L, Silverwood RJ, Guo Y, Ye Z, et al. Association between alcohol and cardiovascular disease: mendelian randomisation analysis based on individual participant data. BMJ 2014;349:g4164.

[67] Lawlor DA, Nordestgaard BG, Benn M, Zuccolo L, Tybjaerg-Hansen A, Davey Smith G. Exploring causal associations between alcohol and coronary heart disease risk factors: findings from a Mendelian randomization study in the Copenhagen General Population Study. Eur Heart J August 2013;34(32):2519–28.

[68] Liu JZ, Tozzi F, Waterworth DM, Pillai SG, Muglia P, Middleton L, et al. Meta-analysis and imputation refines the association of 15q25 with smoking quantity. Nat Genet May 2010;42(5):436–40.

[69] Tobacco and Genetics Consortium. Genome-wide meta-analyses identify multiple loci associated with smoking behavior. Nat Genet May 2010;42(5):441–7.

[70] Åsvold BO, Bjørngaard JH, Carslake D, Gabrielsen ME, Skorpen F, Smith GD, et al. Causal associations of tobacco smoking with cardiovascular risk factors: a Mendelian randomization analysis of the HUNT Study in Norway. Int J Epidemiol October 2014;43(5):1458–70.

[71] Winsløw UC, Rode L, Nordestgaard BG. High tobacco consumption lowers body weight: a Mendelian randomization study of the Copenhagen General Population Study. Int J Epidemiol April 2015;44(2):540–50.

[72] Davey Smith G, Hemani G. Mendelian randomization: genetic anchors for causal inference in epidemiological studies. Hum Mol Genet September 15, 2014;23(R1):R89–98.

[73] McCarthy MI, Abecasis GR, Cardon LR, Goldstein DB, Little J, Ioannidis JPA, et al. Genome-wide association studies for complex traits: consensus, uncertainty and challenges. Nat Rev Genet May 2008;9(5):356–69.

[74] Swerdlow DI, Preiss D, Kuchenbaecker KB, Holmes MV, Engmann JEL, Shah T, et al. HMG-coenzyme A reductase inhibition, type 2 diabetes, and bodyweight: evidence from genetic analysis and randomised trials. [Internet] The Lancet September 2014. [cited 24.09.14]; Available from: http://linkinghub.elsevier.com/retrieve/pii/S0140673614611831.

[75] Naci H, Brugts JJ, Fleurence R, Ades A. Dose-comparative effects of different statins on serum lipid levels: a network meta-analysis of 256,827 individuals in 181 randomized controlled trials. Eur J Prev Cardiol August 1, 2013;20(4):658–70.

Chapter 9

Genetic Susceptibility in Biochemical and Physiological Traits

R.M. Dumitru
St Peter's Hospital, Chertsey, Surrey, United Kingdom

INTRODUCTION

The Human Genome Project has revolutionized cardiovascular medicine. Its completion in 2003 has laid the foundation for groundbreaking discoveries in all of medicine, as genes contribute both to the causes as well as the pathogenesis of human diseases, including cardiovascular disorders. Our knowledge and detailed understanding of certain biochemical and physiological traits grows increasingly faster. The more we know, the easier it becomes to uncover more. New technologies as well as the use of animal models have expedited the field of cardiovascular genetics in the last decade, and progress will predictably grow even faster in the future, even though challenges still remain in advancing from genomic localization to insights in physiology and pathophysiology. Nevertheless, certain pathomechanisms are already relatively well understood. This chapter gives a few examples of genetic susceptibility in biochemical and physiological traits of cardiovascular disease. In addition to these effects on pathophysiology, genes can also influence the response to treatment. Part of this chapter, therefore, examines the opportunities provided by pharmacogenomics for future improvements in treatment of cardiovascular disease.

HYPERTROPHIC CARDIOMYOPATHY

Hypertrophic cardiomyopathy (HCM) is the most common genetic cardiovascular disease, with an estimated prevalence of 0.2% [1]. The actual number of affected patients seen in practice, however, is much lower, as most affected individuals remain unidentified and asymptomatic [2,3]. HCM is a global disease [4] and affects men and women alike, although it seems to be underdiagnosed in women as well as in minorities [3]. Since its first contemporary report in 1958 [5], many different names have been used to describe the same entity, which is mainly due to the very heterogeneous clinical presentation, clinical course, and even prognosis of this disease [3,6]. Names previously used, such as hypertrophic obstructive cardiomyopathy (HOCM) or muscular subaortic stenosis (MSS), are now largely abandoned, and hypertrophic cardiomyopathy (HCM) has become the preferred and generally accepted term [2,3]. The disease is characterized by hypertrophy of the left ventricle without ventricular dilation in the absence of any other condition, cardiac or systemic, that might cause the changes [7]. It is recognized as the most common cause of sudden cardiac death, especially in young individuals and including competitive athletes [8].

HCM is caused by mutations in various genes encoding different protein components of the cardiac sarcomere, the basic functional unit of myocytes (see Fig. 9.1). To date, hundreds of different mutations have already been identified. Different publications state different numbers of discovered genes and mutations associated with HCM, ranging as high as 1400 mutations [9] on 13 separate genes [10]. In a previous study by Van Driest et al., about 40% of referred, unrelated patients with HCM were found to have any one of the known sarcomeric mutations. None of the known causative genotypes were identified in the remaining 60% [11]. HCM is usually inherited as a Mendelian trait in an autosomal dominant fashion [3]. Many patients, however, have no family history for HCM. While part of the reason might be an incomplete family history, or the fact that some mutation carriers remain asymptomatic throughout their life, some patients may have sporadic, de novo mutations [9]. The two genes most commonly associated with HCM are MYH7, encoding the β-myosin heavy chain, which was also the first identified gene in HCM, and MYBPC3, which encodes myosin-binding protein C [12].

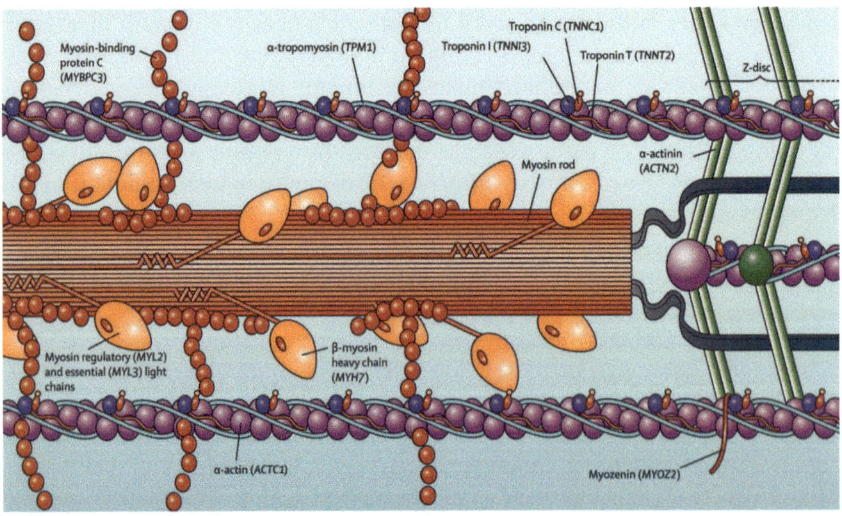

FIGURE 9.1 **The cardiac sarcomere.** *Adapted and reprinted from Maron BJ, et al. Hypertrophic cardiomyopathy. The Lancet. 381(9862):242–55. Copyright (2013), with permission from Elsevier.*

Mutations in these two genes have been found to account for up to 80% of all HCM cases with identified genetic mutations [13]. Roughly 10% are caused by mutations in TNNT2, encoding troponin T, or in TNNI3, encoding troponin I [13]. Other associated genes accounting for much fewer cases include TPM1, encoding tropomyosin α-1 chain, TTN, encoding titin, also called connectin, MYL2, which encodes regulatory myosin light chain 2, MYL3, encoding the essential myosin light chain 3, ACTC1, encoding cardiac alpha actin, and MYH6, encoding cardiac α-myosin heavy chain 6 [3,13]. Different mutations cause different phenotypes in HCM. This even holds true for members of the same family carrying the same mutation in the same gene [3].

The affected proteins are all part of the cardiac sarcomere, the basic contractile unit of myocytes, and therefore are important for the structure and proper function of the myocardium as a whole. Myocytes are made up of myofibril bundles containing myofilaments, which themselves are composed of repeating units called sarcomeres. These microanatomical structures largely consist of thick and thin filaments, whose calcium-dependent conformational changes lead to an intertwining movement that shortens or lengthens the sarcomere and, ultimately, leads to contraction or relaxation of the myocyte [14]. Two further proteins, titin and nebulin, contribute to the structure and stability of sarcomeres. Thick filaments consist mainly of myosin, the α- and β-myosin heavy chains, as well as the regulatory and essential myosin light chains, while the other component of thick filaments is myosin-binding protein C. Thin filaments comprise tropomyosin, actin, and the troponin complex of troponins C, I, and T. Titin is a protein connecting the thick myosin filaments to the so-called Z-disc [15]. As mentioned before, the majority of HCM mutations occur in genes encoding components of the thick filaments; particularly β-myosin heavy chain and myosin-binding protein C. Mutations in genes encoding sarcomere proteins produce defective sarcomeres, and it seems that different mutations may correlate with different clinical expressions, such as age of onset or clinical symptoms. Histopathologically, myofilaments in HCM are often found to be in disarray, aligned in a chaotic and dysfunctional pattern at unusual angles [10]. Many questions regarding the exact mechanisms of how these dysfunctional proteins lead to the many different phenotypes of HCM, however, still remain to be answered.

The paramount feature of HCM is hypertrophy of mainly the left ventricle, clinically usually diagnosed by noninvasive cardiac imaging, such as echocardiography or, increasingly, cardiovascular magnetic resonance imaging (CMR) [3,16]. Various patterns of asymmetric ventricular hypertrophy are found in HCM. Structural and morphological abnormalities of the mitral valve (MV), in the sense of an elongated shape and enlarged size, are also found in a number of HCM patients [3]. Maron et al. [3] emphasized the importance of distinguishing between obstructive and nonobstructive HCM, determined by the presence or absence of a measureable gradient in the left ventricular outflow tract (LVOT).

This subaortic gradient reflects true mechanical impedance to left ventricular (LV) outflow, resulting in increased intraventricular pressure with long-term effects on the LV, increasing myocardial wall stress and oxygen demand, myocardial ischemia, and ultimately myocardial fibrosis [17]. LVOT obstruction also seems to be an independent risk factor of heart failure progression [18]. In most cases, LVOT obstruction is caused by systolic anterior motion (SAM) of the mitral valve [3], which represents a paradoxical motion of mostly the anterior MV leaflet toward the LVOT during systole [19]. SAM is also responsible for concomitant MV regurgitation, commonly of a mild to moderate degree, which is believed to be due to incomplete leaflet apposition [3]. LVOT obstruction is generally dynamic, meaning that most patients do not demonstrate the obstruction under resting conditions, but can be provoked by certain physiological situations, maneuvers and interventions, such as stress through tachycardia, Valsalva maneuver, or hypovolemia [20].

In nonobstructive HCM, symptoms are largely due to diastolic dysfunction, ie, abnormal relaxation and filling of the ventricle in diastole [20], in the presence of preserved systolic function. The filling phase is prolonged, leading to a compensatory increased contribution of the atrial systole to overall filling, and increased left atrial and left ventricular end-diastolic pressures [3]. Diastolic dysfunction has been shown to be present in approximately 80% of HCM patients [20], and while it evidently seems to be based on dysfunctional sarcomeres, the detailed pathomechanisms are not yet fully understood.

Besides LVOT obstruction and diastolic dysfunction, microvascular dysfunction with resulting myocardial ischemia is another common feature of HCM and is based on several factors, including reduced vascular density relative to LV hypertrophy, myocardial fibrosis, diastolic dysfunction, and myocyte disarray [20]. This predisposes patients to exertional angina, syncope, and arrhythmic events.

All in all, the effects of genetic mutations in HCM on physiological and biochemical traits are manifold, and many more, undoubtedly, will be discovered in the future.

CHANNELOPATHIES

One area of cardiovascular diseases, in which genetic influences on physiological and biochemical traits are relatively well understood, is that of channelopathies, arrhythmic disorders caused by defects in cardiac ion channels regulating the cardiac action potential. These channelopathies may be benign and never recognized in asymptomatic individuals, they may predispose to thrombus formation leading to embolus or stroke, or they may cause ventricular arrhythmia, cardiogenic shock, and sudden cardiac death (SCD), which is often enough the first manifestation of the disease. This is all the more associated with great grief, as it may affect individuals in apparently good health, including infants, children, or young adults.

Long-QT Syndrome

Long-QT Syndrome (LQTS) comprises a group of heterogeneous genetic disorders sharing the trait of QT interval prolongation on resting ECG, which represents the time required by the cardiac ventricles to repolarize, and an increased risk of LQTS-associated cardiac events such as syncope, seizures, or SCD, commonly in the context of physical or emotional stress [21]. LQTS represents a leading cause of SCD in young individuals, particularly those with structurally normal hearts [22], as well as a possible cause of sudden infant death syndrome [23].

Classically, LQTS was divided into the autosomal dominant Romano–Ward syndrome (RWS) with an estimated prevalence between 1:2000 and 1:5000, presenting solely with cardiac symptoms, and the extremely rare autosomal-recessive Jervell and Lange-Nielsen syndrome (JLNS), estimated to affect about one in 4 million individuals and causing a cardioauditory syndrome of malignant LQTS and congenital bilateral sensorineural deafness [24]. In reality, however, both the genotypes and phenotypes of numerous mutations underlying autosomal dominant LQTS differ, and as this may also imply different, possibly even contrary treatment strategies, the term Romano–Ward syndrome is less frequently used in newer literature.

While hundreds of different mutations in at least 16 different genes have been found, so far, to be either causative of or at least associated with LQTS (see Table 9.1), three main genes, the so-called "major LQTS genes," are believed to be responsible for approximately 75% of all diagnosed LQTS. These are KCNQ1 (LQT1) mapped to chromosome 11p15.5 [25], KCNH2 (LQT2) on chromosome 7q35-36, and SCN5A (LQT3) on chromosome 3p21-24 [26].

Major Long-QT Genes

KCNQ1, or LQT1, encodes the α-subunit of $K_V7.1$ (K_VLQT1), a voltage-gated potassium channel (K_V) responsible for the slowly activating outward current I_{Ks} in cardiac repolarization [27]. Physiologically, I_{Ks} is increased by sympathetic, β-adrenergic stimulation via PKA-dependent phosphorylation [28], hereby shortening cardiac repolarization, represented by the QT-interval on ECG, when heart rate increases. Defects of $K_V7.1$, like the ones caused by KCNQ1 mutations in LQTS, impair the physiologically increased I_{Ks} during tachycardia and prevent the QT interval from shortening appropriately, potentially leading to arrhythmia, torsades de pointes, and possibly SCD. The connection between sympathetic stimulation, tachycardia, and I_{Ks} increase may explain why syncope and SCD in LQTS are usually precipitated by physical or psychological stress. Cardiac events during swimming are also most frequently associated with mutations in KCNQ1 [27].

The second major LQTS gene is KCNH2, also called hERG (human *ether-à-go-go* related gene) or LQT2, which encodes the α-subunit of $K_V11.1$, a voltage-gated potassium channel conducting the rapidly activating outward

TABLE 9.1 Long QT Genes

LQT Classification	Gene	Locus	Encodes	Effect
LQT1	KCNQ1	11p15.5	α-subunit of I_{Ks} potassium channel, $K_V7.1$	Reduces I_{Ks}
LQT2	KCNH2	7q35-36	α-subunit of I_{Kr} potassium channel, $K_V11.1$	Reduces I_{Kr}
LQT3	SCN5A	3p21-24	α-subunit of cardiac sodium channel, $Na_V1.5$	Increases I_{Na}
(LQT4)	ANK2	4q25-27	Ankyrin-B	Dysfunctional ion channel
LQT5	KCNE1	21q22.1	β-subunit MinK of I_{Ks} potassium channel	Reduces I_{Ks}
LQT6	KCNE2	21q22.1	β-subunit MiRP1 of I_{Kr} potassium channel	Reduces I_{Kr}
(LQT7)	KCNJ2	17q23	Inward rectifier potassium channel Kir2.1	Reduces I_{K1}
LQT8	CACNA1C	12p13.3	Voltage-gated L-type calcium channel $Ca_V1.2$	Increases I_{Ca}
LQT9	CAV3	3p25	Caveolin-3	Increases I_{Na}
LQT10	SCN4B	11q23.3	β4-subunit of cardiac sodium channel, $Na_V1.5$	Increases I_{Na}
LQT11	AKAP9	7q21-22	Yotiao	Reduces I_{Ks}
LQT12	SNTA1	20q11.2	Syntrophin-α 1	Increases I_{Na}
LQT13	KCNJ5	11q24.3	Subunit Kir3.4 of inwardly-rectifying K^+ channel	Reduces I_{KACh}

rectifying current I_{Kr} during cardiac repolarization [29]. Frequently, the channel itself is simply called HERG, although new nomenclature suggests $K_V11.1$ [30]. Both I_{Ks} (see previous) and I_{Kr} are components of the delayed rectifier current I_K, a major parameter in phase 3 of cardiac repolarization. Arrhythmic cardiac events occurring during the postpartum period are often due to mutations in KCNH2. Furthermore, these events may be brought forth by auditory triggers [27].

The third major gene associated with LQTS is SCN5A, or LQT3. This gene on chromosome 3 encodes the α-subunit of the cardiac sodium channel $Na_V1.5$, responsible for conducting the depolarizing sodium current I_{Na}. Mutation in SCN5A increases the sodium inward current ("gain of function"), thus prolonging action potential duration [31] and increasing the risk of possible ventricular arrhythmia. Cardiac events occurring while the patient is asleep or resting are most common in LQT3 [27].

Some patients have multiple mutations in these genes. They usually become symptomatic at a younger age and with increased severity [32].

Minor Long-QT Genes

A group of 8–12 other genes, depending on the literature source, labeled the "minor LQTS genes," make up roughly 5% of all LQTS cases [33]. Two of these rare forms are caused by mutations in auxiliary subunits of two different potassium channels. The gene KCNE1 encodes the β-subunit MinK, which coassembles with the aforementioned K_VLQT1 α-subunits to form I_{Ks} [34]. Mutations in this gene suppress I_{Ks} and cause the LQTS subtype LQT5. LQT6, on the other hand, is caused by dysfunctional β-subunits of the I_{Kr} potassium channel, labeled MiRP and encoded by KCNE2 [35]. Mutations in KCNE2 alter I_{Kr} function, causing the channels to open slowly and close rapidly, which results in a diminished potassium current. This genetically based alteration often remains clinically silent until additional factors, such as certain drugs, further alter the channel function leading to arrhythmia [35].

The auxiliary β-subunit of the cardiac sodium channel $Na_V1.5$ is encoded by SCN4B. Mutations in this gene have been shown to increase the late sodium current and lead to an LQTS phenotype similar to that of LQT3 [36], classified LQT10.

In 2010 Yang et al. [37] further identified KCNJ5 as a susceptibility gene for congenital LQTS (LQT13) by studying consecutive generations of a Chinese family with seemingly inherited LQTS. The gene encodes the subunit Kir3.4 of a G-protein-coupled potassium channel responsible for the inwardly rectifying current I_{KACh}. Previous data had shown that I_{KACh}, whose role in ventricular repolarization might be greater than formerly believed, was obscured by active potassium current I_{K1} [38]. All mutation carriers in this study, apart from the youngest child, experienced recurrent palpitations, and 10 of the 12 mutation-positive family members had recurrent syncopal episodes. Although only three out of the 12 mutations carriers exhibited QTc prolongation greater than 480 ms, five family members, including the three with prolonged QTc, had concomitant torsades de pointes, atrial tachycardia, or atrial fibrillation. The youngest child, whose blood sample was not obtained, suffered from recurrent seizures and syncopal episodes before its death only 7 days after birth. No cause of death was stated in the publication. The authors conclusively suggested that loss-of-function mutation in KCNJ5 affecting Kir3.4 results in loss of I_{KACh} channel function, leading to cardiac arrhythmia including possible LQTS.

Mutations in the gene ANK2, also called ANKB, located on human chromosome 4q25-27, have been found to cause another subtype of LQTS [39], which was the first discovered example of an LQTS form not to result from ion channel mutations. Some sources in literature add ANK2 to the list of minor LQTS-susceptibility genes, while others describe it as a separate entity or add it to a group of three "atypical LQTS genes" or "multisystem syndromic disorders associated with QT interval prolongation." Previously labeled LQT4, it is now preferably referred to as "sick sinus syndrome with bradycardia," or ankyrin-B syndrome [33].

The gene encodes ankyrin-B (also termed ankyrin 2), a member of the ankyrin family further including ankyrin-R (ankyrin 1) and ankyrin-G (ankyrin 3), which have been found in a number of different human tissues and cells, both excitable and nonexcitable, including erythrocytes, neurons, eyes, skeletal muscles, and myocytes [40]. These polypeptides, whose name derives from the Greek word *ankyra* (ἄγκυρα), translating to *anchor* in English, are an important part of the cytoskeleton, linking certain membrane domains such as ion channels and transporters to the cytoplasmic spectrin/actin network [41]. It is therefore easy to understand their important role in providing cell stability and motility as well as ensuring the proper function of specialized membrane proteins. Mutations in ANK2, leading to dysfunctional ankyrin-B, alter the functional organization and ion balance of myocytes and other affected cells through downregulation of Na^+/Ca^{2+} exchanger (NCX), Na+/K+ATPase (NKA), and inositol triphosphate (IP3) receptor. This results in modification of intracellular Ca^{2+} homeostasis with elevated intracellular Ca^{2+} levels and destabilization of the myocyte membrane, rendering the myocardium vulnerable to cardiac arrhythmias [40]. While ANK2 mutations are a relatively rare cause of LQTS, they also predispose to other clinical symptoms, given the role and presence of ankyrin-B in different cell types, which led to the label of ankyrin-B syndrome [42]. Carriers of ANK2 mutations may display other types of cardiac arrhythmias including catecholaminergic polymorphic ventricular tachycardia (CPVT), AV block, idiopathic ventricular fibrillation, and, most notably, sinoatrial node (SAN) dysfunction [40] with bradycardia or junctional escape rhythms, as intracellular Ca^{2+} homeostasis has been shown to be an essential factor in SAN pacemaker activity [43].

In addition to predisposing to cardiac arrhythmias such as LQTS, a study in 2010 demonstrated that faulty ankyrin-B may be a possible underlying cause for developing type 2 diabetes mellitus [44], which itself is a risk factor for cardiovascular disease, while a number of studies further suggest that ankyrin-B may play a role in structural as well as electrical remodeling of the myocardium post MI [45], providing another example of genes' effects on cardiovascular pathophysiology.

Examples of further genes, in addition to ANK2, which are known to cause LQTS without dysfunctional ion channels, include CAV3, AKAP9, and SNTA1. CAV3 encodes caveolin-3 [46], an integral component as an anchoring protein in caveolae and important for vesicular trafficking and signal transduction [47]. Mutations in CAV3 are associated with LQTS (LQT9) by causing an increased late sodium current, late I_{Na}. Further data from Vaidyanathan et al. in 2013 also suggest that CAV3 mutations alter the biochemical properties of Kir2.1, the inward rectifier potassium channel responsible for the potassium current I_{K1}, which adds to the arrhythmogenic effect of LQT9 [48]. However, one study from 2013 [46] questions the role of CAV3 in LQTS, claiming that no association between LQTS and isolated CAV3 mutations has ever been found.

AKAP9 encodes the protein yotiao, which is directly involved in phosphorylation of the potassium channel α-subunit $K_V7.1$ and regulation of the slow outward potassium current I_{Ks} [49]. Certain mutations in AKAP9 negatively influence interaction between yotiao and $K_V7.1$ and reduce cAMP-controlled phosphorylation of the ion channel, thus impairing its response to beta-adrenergic stimulation via cAMP. The consequence is a prolonged action potential with increased susceptibility to QT prolongation [50]. LQTS on the basis of AKAP9 mutation has been labeled LQT11.

The minor LQTS gene SNTA1 encodes syntrophin-alpha 1, a member of the syntrophin family. Alpha-1, contrary to the other syntrophins, beta-1 and beta-2, which are both expressed more widely, is found largely in heart and skeletal myocytes [51], where it acts as a cytoplasmic membrane scaffolding protein. As such, it interacts directly with the pore-forming α-subunit $Na_V1.5$ of the cardiac sodium channel, encoded by SCN5A. Furthermore, syntrophin-alpha 1 is part of a macromolecular protein complex together with neuronal nitric-oxidase synthase-1 (NOS-1) in the heart and the cardiac isoform of plasma membrane Ca2+/calmodulin-dependent ATPase (PMCA4b), which is a potent inhibitor of NOS [52]. Mutations in SNTA1 result in disrupted binding to PMCA4b and reduced inhibition of NOS, which directly leads to increased nitrosylation of $Na_V1.5$. The effect is an increased peak and late sodium current I_{Na}, which may cause a form of LQTS, labeled LQT12, phenotypically similar to LQT3 caused by SCN5A mutation [53,54].

Jervell and Lange-Nielsen Syndrome

As briefly mentioned before, the autosomal recessive form of LQTS is called Jervell and Lange-Nielsen syndrome (JLNS). This rare subtype is caused by homozygous or compound heterozygous mutations of either one of the two genes KCNQ1 and KCNE1, encoding the α- and the β-subunit, respectively, of the potassium channel responsible for the slow component of delayed rectifying current I_{Ks} [55]. About 90% of mutations are found in KCNQ1, which is also associated with a higher risk for arrhythmic events compared to mutations in KCNE1 [56]. Heterozygous mutations in KCNQ1 may lead to LQT1, the most common form of autosomal dominant LQTS, whereas heterozygous KCNQ1 mutations may lead to LQT5, both detailed earlier in this chapter. While ventricular arrhythmia on the basis of QT prolongation with a high risk for SCD is seen in both forms, the major clinical difference between JLNS and the autosomal dominant LQTS forms is the characteristic congenital sensorineural deafness in the former. The bilateral hearing loss has been attributed to absence of functional KCNQ1-KCNE1 pores in the cochlea [57], which have been suggested to be key components for K+ production of endolymph, the essential fluid in the inner ear [58]. One study from 2006 also showed marked differences in severity between JLNS and other LQTS forms [56], and some authors describe JLNS as one of the most severe forms of LQTS altogether [55,56]. Of the 186 patients included in this study, 160 (86%) became symptomatic, defined as having had at least one syncopal episode, aborted cardiac arrest, or SCD, the majority before turning 18 years old. For 15% of the patients, the first cardiac event occurred within the first year of life. In contrast to this, patients with autosomal-dominant LQTS forms tend to become symptomatic much later in life. The main triggers for cardiac events in this particular study were physical exercise and emotional stress, responsible for 95% of all events.

Andersen–Tawil Syndrome and Timothy Syndrome

"Conventional" LQTS classifications further include two rare multisystem disorders, Andersen–Tawil syndrome and Timothy syndrome.

Andersen–Tawil syndrome (ATS) is caused, in the majority of cases [59], by mutations in KCNJ2, which encodes the potassium channel α-subunit Kir2.1 [60]. The channel regulates the inward rectifying current I_{K1}, important for setting resting membrane potential in both skeletal and cardiac muscle cells. Inheritance of ATS is autosomal dominant, although a number of cases have been reported to be sporadic [61] with variable penetrance. Characteristic for ATS is the clinical triad of periodic paralysis, ventricular arrhythmia, and certain physical dysmorphic features [59], such as short stature, low-set ears, cleft palate, clinodactyly, brachydactyly, or syndactyly, and micrognathia [60,62]. Not all patients diagnosed with ATS exhibit all of the typical features, however, as clinical manifestations are quite variable. It is suspected that different KCNJ2 mutations, of which more than 20 are known so far [63], may lead to different phenotypes [60]. While the importance of I_{K1} for membrane potential in skeletal and cardiac myocytes helps explain the symptoms of arrhythmia and paralysis, its role in developing the noted morphological features is not yet understood.

ATS is occasionally still referred to as type 7 LQTS, or LQT7, due to findings of extremely prolonged QTc times on resting ECG. Later studies, however, claimed that initial measurements of QT interval times wrongly included the U-waves, which may be very prominent in ATS [63]. Other ECG characteristics in ATS include ventricular ectopy, bigeminy, and polymorphic or bidirectional ventricular tachycardia [33].

ATS caused by KCNJ2 has been labeled type 1 ATS, or ATS1, as not all cases diagnosed with ATS have underlying KCNJ2 mutations [59]. Zhang et al. [63] suggested that the term ATS1 should now be used instead of LQT7, for distinct ECG changes including prominent U-waves, normal QTc, and characteristic physical features distinguish ATS from actual long-QT syndrome.

Similar to ATS, Timothy syndrome (TS) is a multisystem disorder with QT prolongation and increased risk of cardiac events, also labeled LQT8. Together with JLNS, Timothy TS is considered to be the most malignant LQTS variant [64]. Interestingly, TS seems to be caused by de novo mutations in the affected gene CACNA1C, as opposed to Mendelian inheritance observed in other LQTS variants [65]. Two missense mutations in CACNA1C on chromosome 12p13.3 have been discovered to date, both leading to the same dysfunction in the L-type calcium channel $Ca_V1.2$ [66]. CACNA1C encodes the pore-forming α1-subunit of this channel, which is expressed in different tissues such as the heart, smooth muscle, GI tract, brain, lungs, immune system, and pituitary and adrenal glands [67]. It was also found in developing digits and teeth [65]. Physiologically, $Ca_V1.2$, in response to membrane depolarization, permits Ca^{2+} entry into the cell, where it mediates multiple cell functions like contraction, secretion, and transcription [68]. Mutations in CACNA1C cause impaired voltage-dependent *in*activation of $Ca_V1.2$, which results in increased Ca^{2+} influx and intracellular Ca^{2+} overload in various cell types [65,69]. In the heart, Ca^{2+} influx through $Ca_V1.2$ forms the plateau phase of cardiac action potential, electrically balanced with K^+ efflux through delayed rectifying K^+ channels. The gain-of-function mutation in TS, therefore, causes considerable prolongation of cardiac action potential, which makes the heart susceptible for arrhythmic events and explains the severe QT interval prolongation observed in affected patients [65]. At the same time, the ubiquity of $Ca_V1.2$ and the importance of Ca^{2+} for intracellular processes account for the very broad spectrum of clinical features, including syndactyly, autism spectrum disorder, intermittent hypoglycemia, and immune deficiency [64,65,70]. Some cases further exhibit dysmorphic facial features such as small or misplaced teeth, baldness at birth, flat nasal bridge, low-set ears, and a small upper jaw. Congenital heart disease including patent ductus arteriosus or foramen ovale, ventricular septal defect, and tetralogy of Fallot (TOF) have also been described in a number of affected children [65]. Many of them have language, motor, and cognitive development delays. The most serious traits, however, remain cardiac arrhythmias. In the publication by Splawski et al. [65], 12 of the included 17 children experienced life-threatening cardiac events, 10 of whom died at a mean age of 2.5 years.

Catecholaminergic Polymorphic Ventricular Tachycardia

Catecholaminergic polymorphic ventricular tachycardia (CPVT) is, similar to long-QT syndrome (LQTS), a genetic arrhythmogenic disorder that clinically manifests with cardiac syncope or SCD in patients with structurally normal hearts [71]. As these cardiac events also typically occur during physical exertion or emotional stress, CPVT phenotypically mimics LQTS, particularly LQT1 [72]. Unlike in LQTS, however, the QT interval, as well as all other aspects of the resting 12-lead ECG, is in the majority of CPVT cases completely normal. The most helpful diagnostic test in suspected CPVT has proven to be the stress test, either as a treadmill test or via infusion of catecholamines like the synthetic isoproterenol, where exercise-induced ventricular ectopics occur far more commonly than in patients with LQTS, or with any other cardiac diagnosis [73]. Furthermore, bidirectional ventricular tachycardia has been called the hallmark feature of CPVT, although it is important to note that it is not present in all cases.

In approximately two-thirds of affected patients, autosomal dominant mutations in the gene RYR2, encoding the ryanodine receptor and calcium release channel RyR2, are the underlying cause (CPVT1) [74]. Ryanodine receptors are located in the membrane of the sarcoplasmic reticulum (SR), a cell organelle storing intracellular calcium, with the type 2-receptor subtype, RyR2, being highly expressed in cardiac myocytes. In the normal cardiac cycle, depolarization of the cell membrane leads to opening of voltage-gate calcium channels and calcium influx into the cell. While the amount of inflowing Ca^{2+} is not enough to induce muscle contraction, it is important for the release of larger amounts of Ca^{2+} from the SR, mediated by RyR2 [75]. For this process, RyR2 must interact with further regulatory proteins such as calmodulin, calsequestrin, protein kinase A (PKA) and others, forming a macromolecular complex [76]. One of these proteins is FKBP12.6, an FK506 binding protein [77]. It stabilizes the closed state of RyR2, hereby preventing "leaking" of Ca^{2+} from the SR into the cytosol in the cell's inactive state [78]. Phosphorylation of RyR2 via protein kinase A, initiated by cAMP following catecholaminergic activation of β-adrenergic receptors, weakens the affinity of FKBP12.6 to RyR2, allowing the channel to open, and increasing the response to Ca^{2+}-dependent activation.

Many different mutations in RYR2 have been described to date [79]. In 2003 Wehrens et al. [80] suggested that mutations in RYR2 reduce the affinity between RyR2 and FKBP12.6, causing a "leaky channel" with aberrant Ca^{2+} release during diastole, intracellular calcium overload, and an excessive response to sympathetic nervous system stimulation, triggering ventricular arrhythmia. In contrast to this, another study showed no abnormalities in RYR2/FKBP12.6 interaction and intracellular Ca^{2+} levels in resting cardiomyocytes. The authors did find, however, augmented Ca^{2+} release following RyR-agonist or β-agonist stimulation [81]. Jiang et al., on the other hand, had a different theory altogether [82]. They suggested that RyR2 mutations increased the channel's sensitivity to activation by luminal calcium. Although the underlying mechanism for this theory was not known, the authors suspected a yet-to-be-defined luminal Ca^{2+} sensor.

While the exact pathophysiology underlying RyR2 mutations causing CPVT remains controversial, mutations in another gene causing the same disorder have been identified. This less-frequent subtype, labeled CPVT2, is based on autosomal recessive mutations in CASQ2, encoding cardiac calsequestrin [83]. As briefly mentioned before, this protein is part of the large quaternary complex regulating intracellular calcium levels. Viatchenko-Karpinski et al. [84] suggested that mutant calsequestrin impairs Ca^{2+} storing in the sarcoplasmic reticulum and alters the response of the Ca^{2+}-release complex to luminal Ca^{2+}.

The ultimate effect of either RYR2 or CASQ2 mutations is the same—a heritable arrhythmia syndrome that closely mimics LQT1 without the QT prolongation, and clinically manifests with cardiac events, especially in the young, following physical exertion or emotional stress. Mortality rates for CPVT are much higher compared with LQTS, and may be as high as 30–50% by the age of 50 years.

Brugada Syndrome

The third major heritable arrhythmia syndrome with a high risk for sudden cardiac death, especially in young males, is Brugada syndrome (BrS), first mentioned as a new clinical entity in 1992, when the brothers Pedro and Josep Brugada described the cases of eight patients with a family history for sudden death, structurally normal hearts, and a distinct electrocardiographic pattern [85]. Its clinical presentation is heterogeneous. Possible symptoms range from dizziness or palpitations, to atrial fibrillation, syncope, ventricular arrhythmia, or sudden cardiac death [86], while other affected individuals remain asymptomatic throughout their lives [87]. Arrhythmic cardiac events may occur at any age, although the peak lies around 40 years old [88].

Diagnosis is usually made based on signature ECG changes, commonly divided into types I–III [89] (see Fig. 9.2). The hallmark characteristics of Type I are a J-wave amplitude of 2 mm or more, gradually descending terminal ST segment, coved-type ST-T configuration and negative T-wave [89], all in the right precordial leads V_1–V_3.

If Type I ECG is present with other specific findings such as an episode of ventricular arrhythmia, a positive family history for SCD before the age of 45 years, or a syncopal event, BrS is very likely [89]. If Type II or III ECGs are present with a typical history, it may be possible to induce Type I ECG following a challenge with a sodium channel blocker such as ajmaline, again making BrS the likely diagnosis. These characteristic ECG changes are occasionally accompanied by atypical right bundle branch block, PQ prolongation, axis deviation, or QRS widening [90] (see Fig. 9.3). Initially calling it "syndrome of right bundle branch block, persistent ST segment elevation, and sudden death," the

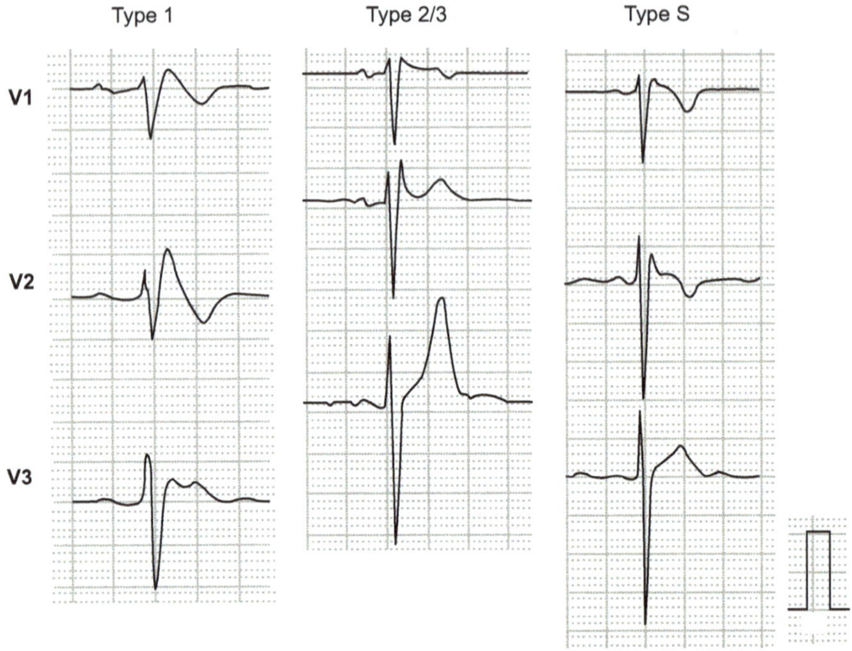

FIGURE 9.2 **Typical Brugada ECG changes.** *Reprinted from Nishizaki M, et al. Classification and assessment of computerized diagnostic criteria for Brugada-type electrocardiograms. Heart Rhythm. 7(11):1660–66. Copyright (2010), with permission from Elsevier.*

Type 1 : (coved-type ST-segment elevation)

① J point ≥0.2mV

② STmax >STmax40 >STmax80

③ T wave: under or on the isoelectric line

④ 0.4mV ≥STmax-STmax40

Type 2/3 : (saddleback-type ST-segment elevation)

① J point ≥0.2mV

② J point >STmin

③ Tpeak >STmin >0mV

Type S : (mild coved-type ST-segment elevation)

① 0.2mV >J point ≥0.1mV

②③: same criteria as Type 1

④ 0.4mV ≥STmax-STmax40 ≥0.04mV

⑤ STmax40-STmax80 ≥0.04mV

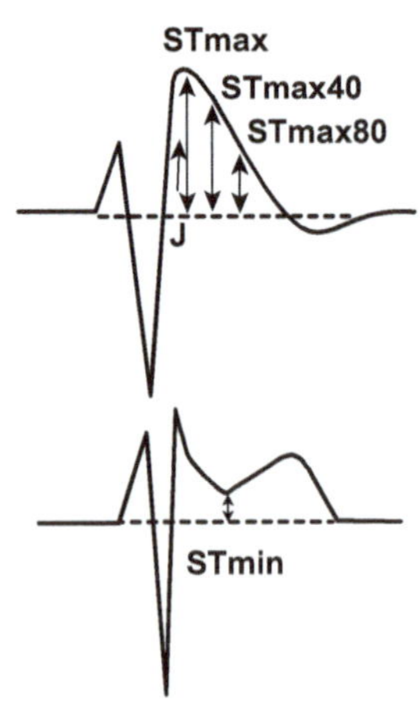

FIGURE 9.3 **Diagnostic criteria for Brugada types 1, 2/3, and S.** *Reprinted from Nishizaki M, et al. Classification and assessment of computerized diagnostic criteria for Brugada-type electrocardiograms. Heart Rhythm. 7(11):1660–66. Copyright (2010), with permission from Elsevier.*

two authors of the initial publication from 1992 later accepted the suggestion of other scientists to simply and more practically refer to it as Brugada syndrome.

The first genetic defect associated with BrS was identified in 1998, when mutations in SCN5A were discovered in patients displaying the clinical BrS characteristics. To date, over 300 mutations in SCN5A alone have been published [91] in connection with BrS, while as many as 12 different genes in total are suspected to be involved with this disease [92]. SCN5A encodes the α-subunit of the voltage-gated cardiac sodium channel $Na_V1.5$. Mutations in this gene, suspected to account for up to 30% of all known BrS cases and designated BS1 [93], may lead either to reduced density of the sodium channel through frameshift or deletion mutations, or to altered channel functions such as activation, inactivation, or reactivation through missense mutations. Both ultimately lead to a reduced I_{Na} current, responsible for phases 0 and 1 of the cardiac action potential. The SCN5A mutations leading to LQT3 cause a *gain* of function with an *in*creased I_{Na} current. Certain mutations can, in fact, cause both LQT3 and BrS in the same person [94]. Other suggested mutations leading to BrS include those in glycerol-3-phosphate dehydrogenase 1-like, labeled BS2, which disturbs the trafficking of the cardiac Na^+ channel to the cell surface, hereby reducing the channels' density on the cell surface and leading to I_{Na} decrease [93]. Mutations in CACNA1C (BS3) and CACNB2 (BS4), encoding the α- and β-subunits of the cardiac L-type Ca^{2+} channel, cause a decreased I_{Ca} current, which results in a combined BrS and short QT syndrome [95]. The list of suspected genes contributing to BrS, also including SCN1B, SCN3B, KCNE3, KCNE5, KCND3, and MOG1, might become even longer in the future.

Nonetheless, in a large number of patients diagnosed with BrS, none of the known mutations are found [96]. Hence the pathophysiology of BrS remains controversial. The two main theories are the repolarization-defect theory and the depolarization/conduction theory. The former is based on shifts in ion currents at the end of phase 1 of cardiac action potential, namely a decrease of inward positive current or an increase of outward positive current, with differing transient outward currents (I_{to}) between the epicardium and other layers of the right ventricular myocardium. This leads to the typical ST changes on ECG and susceptibility to ventricular arrhythmia originating in the right ventricular epicardium [97]. The depolarization theory states that slow conduction and activation delays, mainly in the right ventricular outflow tract (RVOT), pose the crucial pathophysiology in BrS. Further studies have produced evidence for each of these two theories, which might not be mutually exclusive. New theories are being developed, and some scientists have tried to unify different theories in order to explain the pathophysiology of BrS. Given the number of seemingly involved genes and total number of associated mutations, however, there might not be only one mechanism underlying BrS. It is, in fact, quite possible that several different physiological changes, maybe even different disease altogether, lead to the similar, only slightly differing clinical and electrocardiographic manifestations that we currently summarize as Brugada syndrome.

CONNECTIVE TISSUE DISORDERS

Marfan Syndrome

Marfan syndrome (MFS) is a systemic connective tissue disorder, which predominantly affects the skeletal, ocular, and cardiovascular systems [98]. It is estimated to affect two to three per 10,000 individuals, seemingly consistent worldwide, and following the same distribution in both genders [99]. Albeit most affected patients inherit the disease in an autosomal dominant fashion, up to one-third of cases occur due to de novo mutations in either parental gamete [100].

The clinical presentation is variable. Common features of the ocular system include myopia and ectopia lentis, a displacement of the lens from its normal location in the center of the pupil [98]. Patients with MFS also have a higher risk for retinal detachment, early cataracts, or glaucoma [99]. Skeletal signs are multifold. Marfan patients tend to be of tall stature with long-bone overgrowth, or dolichostenomelia, arachnodactyly, wrist and thumb signs such as the Walker–Murdoch and the Steinberg signs, joint hypermobility, pes planus, high-arched palate, and many others [99]. The pulmonary system might be affected either by skeletal deformities such as pectus excavatum or carinatum restricting lung expansion, or by widening of the distal airspaces, which predispose patients to spontaneous pneumothoraces. Widening of the dural sac (dural ectasia) is found in two-thirds or more of patients, which can lead to lower back pain, although many patients remain asymptomatic [99]. The entity that generally causes the most worry, however, remains the cardiovascular system. To be more specific, the clinical signs can be divided into those affecting the heart and those affecting the vascular system. Common findings in the heart are prolapse of the mitral or the tricuspid valve, or both, each with or without regurgitation, which can lead to later complications such as cardiomyopathy, heart failure, or ventricular arrhythmia. Calcification of the mitral annulus is found occasionally and is especially remarkable if present before 40 years old. Aneurysms of large arteries such as the main pulmonary artery or the descending thoracic or abdominal aorta are the most common vascular signs. Dilation of the aortic root with its high risk for aortic root dissection is hereby the most life-threatening manifestation. However, these days, short-interval monitoring, early recognition of disease progression, and very high success rates in aortic root replacement, if necessary, ensure a nearly normal life expectancy in patients with MFS compared to unaffected individuals [99].

MFS is most prominently associated with mutations in the gene FBN1 on chromosome 15q21.1, which encodes the connective tissue protein fibrillin-1 [100]. Monomers of this glycoprotein aggregate to form microfibrils, wildly distributed as a structural component in the extracellular matrix (ECM) of various elastic and nonelastic tissues [98]. After the initial discovery of FBN1 mutations in MFS and the attributed role of fibrillin-1 to organ structures, it was believed that these mutations caused abnormal fibrillin, which was then incorporated into microfibrils creating structurally inferior connective tissue with disturbed elastogenesis. This would subsequently lead to impaired mechanical qualities of various tissues and cause the phenotypes typical for Marfan disease.

While this might not be wrong, it seems to represent only one part of MFS pathophysiology, as recent studies have shown that microfibrils may not be essential for the assembly of elastic fibers in fetal development, but may, in fact, be more relevant for maintaining already-established elastic fibers later in life [101]. It was also discovered that MFS pathophysiology, most notably that underlying the feared complication of aortic aneurysms, is a combination of altered mechanical characteristics of certain tissues, increased activation of transforming growth factor beta (TGF-β), and altered cell-matrix connections [102], rather than merely a structural problem. Increasing evidence supports the theory that MFS patients, due to the inferior quality of fibrillin-1 caused by FBN1 mutations underlying this genetic disorder, begin their life with a deficiency of connecting filaments in the ECM. Over time, more and more of the residual connections will vanish, due to homeostatic changes in the ECM and exaggerated elastolysis. Matrix-degrading enzymes like the matrix metalloproteinase 2 (MMP2) and 9 (MMP9) begin the transformation of ECM, followed by infiltration of inflammatory cells, as well as infiltration and proliferation of vascular smooth-muscle cells. The final result is elastic fiber calcification, thickening of the arterial wall, especially of the intima media, and structural collapse of the arterial wall [103]. These vascular pathological changes, initially discovered in mouse models, were later identified in seven human patients diagnosed with MFS.

In addition to the described, rather obvious, structural role of fibrillin-1, there has been increasing evidence about its important regulatory role for cytokines. TGF-β, in particular, has been identified to be a key target for fibrillin-1 and to play a major role in MFS pathogenesis. This multifunctional growth factor, divided into subtypes 1–3, controls a wide spectrum of different cell functions, including proliferation, differentiation, matrix deposition, and apoptosis [104]. Many different cell types secrete a biologically inactive large latent complex (LLC), consisting of TGF-β, the so-called latency-associated peptide (LAP) and the latent TGF-β binding protein (LTBP), into the ECM. Physiologically, fibrillin-1, which is one of different matrix-binding proteins, sequesters LLC and thus limits the activation of TGF-β [105]. Dysfunctional fibrillin-1, as seen in FBN1 mutations in Marfan syndrome, or a reduced level thereof, is not able to control TGF-β activity due to inadequate sequestration, which leads to increased TGF-β activity with dismantling effects on the extracellular matrix [98]. This mechanism was proposed to explain the pulmonary emphysematous changes [106] observed in MFS. Mouse models, frequently used for MFS studies, showed distal airspace enlargement in fibrillin-1-deficient mice, changes similar to the ones seen in human Marfan patients. Lung histology examined by Dyhdalo et al. [106] in these human individuals showed a pattern of distal acinar emphysema in all examined patients. Neptune et al. [107]

studied fibrillin-1-deficient mice and found an abnormal lung development with impaired distal alveolar septation. All these mice showed markedly increased TGF-β levels and a dysregulation of TGF-β activation and signaling, which resulted in apoptotic developing lungs. Giving these fibrillin-1-deficient mice TGF-β-neutralizing antibodies postnatally was sufficient to rescue alveolar septation. Following these findings, it seemed reasonable to suspect that other typical manifestations of MFS were similarly caused by increased TGF-β activity, particularly those that could not be explained by the structural function of fibrillin-1 alone. Ng and coauthors suggested that TGF-β was involved in the pathogenesis of myxomatous thickening of the mitral valve [108], a relatively frequent phenotype in MFS. They studied mitral valves of fibrillin-1-deficient mice and discovered a correlation with cell proliferation, reduced apoptosis, and excessive TGF-β activation. Analogous to Neptune's findings on lung changes, in vivo antagonism of TGF-β through TGF-β antibodies rescued the myxomatous valves in Ng's fibrillin-1-deficient mice.

Loeys–Dietz Syndrome

The important role of TGF-β was highlighted further with the relatively recent discovery that mutations in the TGF-β gene itself are associated with a phenotype very similar to Marfan syndrome. In 2005 Loeys, Dietz, et al. described a new syndrome, later simply called Loeys–Dietz syndrome (LDS), which shared some of the characteristics of MFS [109]. These included chest deformities, scoliosis, joint hypermobility, arched palate, dural ectasia, ascending aorta aneurysms, and dissections, and others. However, long-bone overgrowth in LDS was rather subtle in comparison to MFS, and there was no ectopia lentis. Some new features were described in LDS that were not reported for MFS cases previously, most notably marked arterial tortuosity, aneurysms, and dissections throughout the arterial system, extending beyond the typical locations associated with MFS with an often aggressive course and frequently leading to death in childhood. Other unique characteristics were craniosynostosis, learning disability, easy bruising, translucent skin, cleft palate, and congenital heart disease [109]. None of the patients was a carrier for an FBN1 mutation. More recent literature reviews and new classifications suggested dividing LDS, which presents with a wide spectrum of multisystem features, into four subtypes depending on the underlying mutation [110]. The first two reported genetic causes were found in two genes encoding the TGF-β receptor, namely TGFBR1 on chromosome 9 and TGFBR2 on chromosome 3, and LDS subtypes caused by mutations in either of these two genes were labeled LDS1 and LDS2, respectively. The latter is by some authors synonymously used for the disorder also described as MFS2, short for Marfan syndrome type 2, or "MFS without major ocular signs" [111]. Mutations in the SMAD3 gene on chromosome 13 underlie LDS3, also called "aneurysms-osteoarthritis syndrome." The gene encodes the protein with the fairly unusual name of "Mothers against decapentaplegic homolog 3," a member of the SMAD protein family involved in the TGF-β pathway.

Activated TGF-β receptor type I leads to phosphorylation and thus activation of SMAD2 and SMAD3, which are then translocated to the nucleus as a complex with SMAD4, where they regulate expression of target genes [112]. LDS4, the fourth proposed subtype of LDS, is caused by mutations in the gene encoding transforming growth factor β2 ligand, TGFB2, another important element of the TGF-β signaling pathway. As no specific clinical criteria have been determined, the diagnosis to distinguish between the four subtypes of LDS will ultimately need to be confirmed by a molecular test [110].

While FBN1 mutations in MFS lead to an increase of TGF-β activity and signaling, one would expect a decrease in any of the four mutations leading to each of the four LDS subtypes, a subject that has been controversially discussed. In reality, however, these mutations result indeed in seemingly paradoxically increased TGF-β signaling, which has been shown in different studies to date [113,114]. One hypothesis for this is autocrine or paracrine compensatory upregulation of TGF-β [113], although the exact mechanism is not yet fully understood.

GENETIC DISORDERS LEADING TO ATHEROSCLEROSIS

Familial Hypercholesterolemia

Familial hypercholesterolemia (FH) was the first of now several known monogenic disorders shown to underlie raised plasma cholesterol levels. Carl Miller, a physician at the Oslo County Hospital, demonstrated in the 1930s that the signs of tendon xanthomas, high plasma cholesterol, and premature coronary heart disease were often found grouped together in families. He provided clues for an association between blood lipids and the development of atherosclerosis, as well as delivering the basis of FH [115]. Taking these findings further, Joseph Goldstein and Michael Brown, who shared the Nobel Prize in Physiology or Medicine in 1985 [116], showed that LDL is removed from the blood through LDL receptors, and that FH is a result of mutations in these receptors, leaving them unable to bind and to internalize LDL into the cells [117]. The result is hypercholesterolemia in affected patients, who are at increased risk for premature cardiovascular disease [118]. Over 1600 mutations in the LDLR gene, which transcribes for the LDL receptor, have been identified so far. Other mutations known to cause FH affect genes for apolipoprotein B and proprotein convertase subtilisin/kexin-type 9 (PCSK9) [119]. FH is inherited primarily in autosomal dominant fashion. Heterozygote individuals, carriers of one mutant copy of the LDLR gene, produce approximately half the normal number of LDL receptors; their plasma LDL levels are usually increased two- to threefold [120]. Homozygote individuals, carriers of two mutant copies, have LDL levels of six times or more of normal levels. These patients, who suffer from severe atherosclerosis, usually die from myocardial infarction (MI) during their childhood [121]. Overall, however, mortality from coronary artery disease varies considerably in patients with FH.

This seems to be due to the type of LDLR mutation as well as the influence of the same environmental factors such as nicotine, low activity levels, and diet, which also affect other, non-FH, individuals [122].

While FH is considered a rare disorder, it is believed to be highly underdiagnosed. Prevalence for heterozygous FH was believed to be 1/500 and about 1/1,000,000 for homozygous FH [123]. Newer data, however, suggest that these numbers vastly underestimate the prevalence of FH. According to the Dutch Lipid Clinic Network criteria for diagnosis of FH in adults, which is largely considered the "gold standard," it is now estimated that about in 200 people have definite or, at least, probable FH. This means that over 34 million people might possibly be affected by FH, which is more than double the previously suggested number [124].

Lipoxygenases

It had previously been shown that atherosclerosis as a principal cause of cardiovascular disease is initiated by low-density lipoproteins (LDL), which become entrapped within the arterial wall and are modified through oxidation, leading to production of inflammatory molecules in vascular cells. This initiates a cascade consisting of leukocyte recruitment, mainly macrophages, further oxidation of accumulating lipoprotein, and fibroproliferation [125].

Subsequent studies identified a number of lipid-modifying enzymes within the arterial wall, which are able to generate inflammatory mediators, thus stimulating further plaque growth [126]. Among these enzymes is 5-lipoxygenase (5-LO). It is produced by leukocytes, primarily in macrophages [127], and plays an important part in leukotriene synthesis. It is the rate-limiting enzyme in the conversion of arachidonic acid into leukotrienes [128], which are inflammatory mediators associated with the attraction of monocytes. This suggests that 5-LO might at least contribute to chronic inflammation and thus the progression of atherosclerotic lesions [129].

Mehrabian et al. [130] identified a major locus for atherosclerosis on mouse chromosome 6. In their studies, they created a resistant phenotype designated "CON6" through crossing atherosclerosis-resistant mouse strains with a susceptible strain. The CON6 mice had dramatically reduced lesion formation despite being bred onto a high-LDL background. They further examined the congenic region for potential candidate genes and found that 5-LO mapped adjacent to the linkage peak. They performed a subsequent study, in which they examined the role of 5-LO to atherosclerotic lesion formation, its contribution to inflammation, as well as its synthesis in leukocytes. Their findings suggested that 5-LO does indeed participate in atherogenesis in mouse models, as the enzyme was found in large quantities in macrophage-rich parts of atherosclerotic plaques. The results suggested that alteration or loss of 5-LO function through variations of the 5-LO gene could explain the resistance to developing atherosclerosis in resistant CON6 mice [131].

Taking these results from previous mouse models further, a study by J.H. Dwyer et al. aimed to prove that polymorphisms in the 5-LO promoter affected atherosclerosis in human patients. The genotypes of 470 patients, who were all free of diagnosed cardiovascular disease, yielded 94% carriers of a common allele and 6% carriers of two variant alleles, while other major risk factors for cardiovascular disease were fairly equal in both groups. Carotid intima-media thickness of all participants was then compared, revealing elevated thickness in the carriers of the two variant alleles compared to the carriers of the common allele. The difference remained significant after adjustment for other risk factors. This association was suggestive of 5-LO polymorphisms playing a role in human atherosclerosis.

Furthermore, the group compared a diet–gene interaction in the same study, showing that increased dietary consumption of n-6 fatty acids arachidonic acid and linoleic acid, which is the metabolic precursor of arachidonic acid, led to increased severity of atherosclerosis, however, this effect was only seen in the group carrying the two variant alleles. High intake of n-3 fatty acids, which are found in fish oil, on the other hand, had the opposite effect of blunting the seeming atherogenic effect of the previously described two variant genotypes [132].

These observations put a focus on genes affecting 5LO-synthesis as candidate genes involved in atherosclerosis and thus cardiovascular disease.

Helgadottir et al. [133] found a suggestive linkage peak on chromosome 13 in a linkage study that involved 296 Icelandic families, including over 700 individuals with a history of MI. The group further investigated this particular chromosomal region with microsatellite markers, finding a four-marker single-nucleotide polymorphism haplotype in the gene ALOX5AP, which encodes for 5-LO-activating protein (FLAP), to be associated with a twofold increased risk of developing both MI and stroke in the examined population. In a separate, independent cohort with British individuals, Helgadottir et al. repeated the association of ALOX5AP with MI, this time, however, with a different haplotype than the one they discovered in the initial linkage study in Iceland.

This evidence for the involvement of the lipoxygenase pathway also makes the remaining genes of this inflammatory pathway in atherogenesis attractive candidates for further genetic studies in the future.

Myocyte Enhancer Factor 2A

Aiming to better understand the genetic basis of coronary artery disease (CAD), Wang et al. performed a genome-wide linkage study in 2003 with members of one large family [134]. Of the 21 individuals included, 13 were symptomatic demonstrating a dominant pattern of CAD, including nine patients who developed acute MI. The authors identified a linked region on chromosome 15q26, which was designated *ad*CAD1, as it was the first published autosomal dominant CAD and MI locus, containing at least 93 genes. Among these genes, 43 were known, 50 were hypothetical. The team focused their interest on the myocyte enhancer factor 2A gene (MEF2A), as it had previously been shown to play a role in mouse embryogenesis [135] and was furthermore suspected to play a role in vasculogenesis and vascular morphogenesis due to similarities in expression patterns of MEF2A, vascular endothelial growth factor receptor 2, which is found in endothelial cell precursors, and the Von Willebrand factor (vWF), which is considered an endothelial cell marker. Systematic mutational screening with direct DNA sequence analysis of the entire MEF2A gene identified a 21-base-pair deletion causing deletion of seven amino acids (Δ7aa) of MEF2A in all 10 living family members that were affected by CAD. The amino acid deletion, Δ7aa, was not found in any of the family members expressing normal phenotypes or in anyone from a separate collective of 119 individuals with normal coronary angiograms, leading to the conclusion that the deletion in MEF2A was directly and mainly responsible for CAD in the examined family. Wang et al. hypothesized that Δ7aa led to a conformational change of the MEF2A protein, thus limiting its function as a transcription factor. Immunofluorescence staining showed mutant MEF2A to be retained in the cytoplasm, while wild-type MEF2A was localized in the nucleus, suggesting a possible transportation barrier into the nucleus for mutant MEF2A. Transcription activation assays also showed limited transcription activity of only one-third of mutant MEF2A containing the Δ7aa deletion compared to wild-type MEF2A. These in vitro studies suggested that the deletion Δ7aa affected gene function. The same authors further revealed a strong signal for MEF2A protein within endothelial cells of coronary arteries as well as human umbilical vascular endothelial cells, leading to the conclusion that MEF2A must play an important role in the development and normal function of endothelial cells. The team suspected that a genetic defect in MEF2A could lead to abnormal endothelial function and promote the growth of atherosclerotic plaques. Wang et al. did mention, however, that mutational analysis of 50 sporadic patients with CAD and MI failed to detect any mutations in MEF2A and that no MEF2A mutations have been identified in three other large families with prevalent CAD, leading to their assumption that CAD/MI might be genetically heterogeneous and that MEF2A mutations might, in fact, be a rather rare cause of coronary artery disease. They also did not want to rule out the possibility that a different mutation in a gene unidentified at the time of their studies might also be involved in the development of CAD/MI in the examined family.

A further study was published by Bhagavatula et al. [136], in which the authors identified three new mutations of MEF2A found in four of 207 patients (1.93%) with CAD/MI, while the same or any other MEF2A mutations were not found in 191 control individuals whose coronary angiograms were normal. While the described Δ7aa mutation acted through a dominant-negative mechanism, these three new mutations acted by loss of function. Functional studies performed by the authors indicated that all three of the newly described mutations led to reduced transcription activation activity. Carriers of these three mutations, neither of whom developed MI, were also shown to suffer from less-severe CAD compared to the Δ7aa mutation carriers, as 9 of 13 individuals with the Δ7aa deletion suffered MI. The authors estimated from their data that approximately 2% of patients with CAD carry mutations in MEF2A.

Subsequently, Weng et al. [137] aimed to confirm the previous results in a follow-up study involving 300 patients with premature CAD, but failed to find any causative MEF2A mutations in the cohort. They did, however, find the 21-bp MEF2A deletion in one elderly control subject without CAD.

While the role of MEF2A mutations as a cause for CAD remains controversial, the findings of the previous studies do not rule out the earlier suggested participation of MEF2A in vasculogenesis, meaning that MEF2A variants might indeed play a part in susceptibility for CAD [138]. It might also be plausible to think that a further genetic or environmental factor could be necessary to enhance the role of MEF2A in development of atherosclerosis.

Tangier Disease and Sitosterolemia

Tangier disease (TD), or ATP-binding cassette transporter A1 deficiency, is a rare hereditary although severe form of high-density lipoprotein (HDL) deficiency. It was first described in 1961 by Dr. Donald Fredrickson et al. [139], after physicians at the National Cancer Institute were introduced to a five-year old boy from Tangier Island in Chesapeake Bay, Virginia, USA, with unusually large and orange tonsils, splenomegaly, and lymphadenopathy. The authors soon discovered that the boy's sister had the same symptoms, and that both children as well as both of their parents had very low HDL levels, leading the authors to suspect an autosomal recessive genetic disorder [140]. German scientists Rust et al. successfully assigned TD to the long arm of chromosome 9 (9q31) through genome-wide graphical linkage exclusion [141]. Combining these findings with previous discoveries of genes for two ATP binding cassette (ABC) transporters on chromosome 9, namely ABCA1 and ABC2 [142], led to the conclusion that mutations in ABC1 caused Tangier disease through impaired cholesterol homeostasis and transport [143].

The ABCA1 gene encodes the ATP binding cassette transporter A1, which normally promotes phospholipids and free cholesterol out of peripheral cells and onto lipid-poor apolipoprotein A1 (apoA1), thus forming preβ-HDL, a precursor of HDL. Mutation of ABCA1 therefore leads to reduced apolipoprotein-mediated efflux of lipids from cells with accumulation and deposition of cholesterol esters in different tissues of the body as well as very low plasma levels of HDL [143]. As low HDL has previously been shown to be an important risk factor in development of atherosclerosis and cardiovascular disease [144], Tangier disease, which causes low HDL, predisposes to these pathologies and to premature MI [145].

Since its first description, TD has been diagnosed in about 100 patients worldwide [146]. The clinical features are caused by accumulation of cholesterol esters in different organs, which can lead to hepato- and splenomegaly, thrombocytopenia, anemia, corneal opacities, neuropathy, gastrointestinal disorders, as well as atherosclerosis and its consequences such as premature MI, which has been reported in about one-third of all known TD cases, and stroke. Biochemical characteristics are low HDL, total cholesterol, and apoA-I, and normal to high plasma triglycerides. Almost all children diagnosed with TD were identified by enlarged yellow-orange tonsils, while many of the adult TD patients initially presented with signs of neuropathy [146].

Heterozygote mutation carriers usually are clinically asymptomatic, although their plasma HDL and apoA-I levels lie between those of healthy individuals and homozygote mutation carriers of ABCA1, whose plasma HDL and apoA-I can both reach levels below 5 mg/dL. This is due to the clinical phenotype being inherited in autosomal recessive fashion, hence largely asymptomatic heterozygous individuals, whereas the biochemical phenotype is inherited as an autosomal codominant trait [147,148]. Further studies did show, however, that ABCA1 heterozygotes displayed a higher prevalence for coronary artery disease [149].

While TD is an extremely rare disease, it is conceivable that the diagnosis is sometimes missed despite its imposing clinical and biochemical presentation. Furthermore, studies on TD and the underlying ABCA mutations have contributed greatly to understanding the role and importance of HDL and its atheroprotective mechanisms [150] to the point that HDL and its relation to cardiovascular disease has prompted extensive interest in the development of pharmacological agents that can raise plasma HDL [151].

Another disease based on ABC transporter mutation and affecting the cardiovascular system is sitosterolemia, a rare, autosomal recessive disorder caused by loss-of-function mutations in the ABC transporter genes ABCG5 and ABCG8. The first known case involving two sisters with extensive tendon xanthomas and elevated blood levels of beta-sitosterol was published in 1974 [152]. Beta-sitosterol is a phytosterol, or plant sterol, and thus similar to cholesterol, an animal sterol, in its chemical structure [153]. The genetic defect underlying sitosterolemia was localized on the designated sitosterolemia locus STSL on the short arm of chromosome 2p21 [154] and shown to affect the genes ABCG5 and ABCG8, which are located in close proximity to each other in head-to-head orientation and encode two members of the ABC-transporter family [155,156], sterolin-1 and sterolin-2, respectively. The proteins encoded by those two genes form a heterodimer in the endoplasmic reticulum, and coexpression of both proteins is necessary for their transport onto the apical surface of hepatic and intestinal cells [157]. Physiologically, they interact in order to limit intestinal absorption of sterols and to facilitate their biliary excretion [155,158]. High-cholesterol diets have been shown to increase ABCG5 and ABCG8 mRNA expression in mouse models [159]. Homozygous carriers of mutations in either ABCG5 or ABCG8 hyperabsorb cholesterol as well as plant sterols from the intestine while, at the same time, elimination of sterols into bile is impaired [154], leading to vastly elevated sitosterol plasma concentrations, or sitosterolemia. The patients may suffer from xanthomas, hemolytic anemia, thrombocytopenia, premature atherosclerosis, and coronary artery disease [160]. Interestingly, mutations in the closely neighboring ABCG5 and ABCG8 genes on chromosome 2p21 lead to the same phenotype [155]. It should be noted, however, that some recent studies have challenged the causality of ABCG5 or ABCG8 mutations for premature atherosclerosis in patients suffering from sitosterolemia [161], suspecting the vascular changes to be, at least partially, due to other underlying mechanisms, independent from high plasma phytosterol levels. Future studies will, undoubtedly, clarify this divisiveness.

Even though sitosterolemia is a rare disease, its research has, similar to the studies on Tangier disease with HDL, helped to understand the mechanisms behind sterol absorption and excretion, both physiologically and pathologically.

CARDIOVASCULAR PHARMACOGENETICS AND PHARMACOGENOMICS

Drug therapy plays a crucial role in the treatment of cardiovascular disease. A common observation in different clinical settings is that some patients seem to not only benefit more from certain drugs than other patients but they also seem to suffer from substantially fewer and less dramatic adverse effects. While part of this is certainly due to patient noncompliance, to drug interactions that might be unrecognized, or to organ dysfunction such as kidney disease resulting in reduced drug excretion or liver disease leading to impaired drug activation or elimination, another component of this variability has been shown in a number of studies to be genetically predetermined. Genetic predisposition can affect either or both pharmacodynamic and pharmacokinetic factors of drug therapy. Two of the most obvious effects are drug metabolism and functional variations in genes encoding different drug receptors, although other elements in the downstream signaling pathway might also be affected.

Pharmacogenetics is the study of these genetic influences on the response to drug treatment, or, more precisely, of the influences of single genes on drug therapy, whereas pharmacogenomics is a broader term, relating to the entirety of all genes, the genome, and their influence on drug response. Commonly, however, the two terms are used interchangeably.

Pharmacogenomics is already practically applied in certain fields such as oncology, where the drug trastuzumab, for example, is generally used for treatment of certain types of breast cancer, namely those that are positive for the HER2 receptor. HER2, or human epidermal growth factor receptor 2, physiologically promotes cell growth and differentiation, while its overexpression results in an enhanced reaction to growth factors and malignant growth. Trastuzumab, an anti-HER2 monoclonal antibody, interferes with this pathway by removing the receptor from the cell surface through endocytosis [162]. Patients diagnosed with breast cancer may be tested for this receptor, to determine whether they might be suitable for treatment with trastuzumab, as the drug is usually not offered to patients who are HER2-negative, where it is believed that the side effects of treatment outweigh its benefits.

As for pharmacogenetics of cardiovascular disease, most of the data has been collected from studies related to warfarin, an anticoagulant interfering with synthesis of Vitamin K-dependent clotting factors, and clopidogrel, a thienopyridine antiplatelet agent [163]. Other substances frequently used in the treatment of cardiovascular disease, however, such as aspirin, statins, and β-blockers, have also been shown to be subject to this interindividual variability in efficacy and toxicity.

Pharmacogenetics of Warfarin

Interindividual differences in response to warfarin are caused by both pharmacodynamic and pharmacokinetic factors. This vitamin K antagonist is arguably the most commonly prescribed oral anticoagulant, used for prevention and treatment of thrombosis and thromboembolic disease, although it is believed to still be underused due to one major downside, which is its narrow therapeutic index. If dosed too low, it will be ineffective, whereas a dose that is too high may lead to an overshooting reaction and increase the risk of both major and minor hemorrhage. It is therefore crucial to determine a safe but effective initiation and stabilization dose as well as to adjust the dose regularly to adapt to changes in patients' diet, disease state, or comedication.

Several clotting factors, including the factors II (prothrombin), VII, IX, and X, as well the anticoagulant proteins C and S [164], depend on vitamin K as a cofactor [165] in order to exert their anticoagulant effect. Warfarin inhibits vitamin K epoxide reductase VKORC1, the enzyme responsible for transforming vitamin K epoxide into the cofactor vitamin K hydroquinone [166] (see Fig. 9.4).

Different studies have shown that polymorphisms in the gene encoding VKORC1 account for up to 30% of the observed variability in stabilized warfarin dose [167]. A number of different polymorphisms and haplotypes have been defined, and different classifications are used in literature, most commonly the H haplotype terminology of Rieder et al. [180] and the star terminology introduced by Geisen et al. [181]. Generally, they may be divided into low-dose, intermediate-dose, and high-dose haplotype groups.

One polymorphism in the promoter region of VKORC1, called G3673A, or 1639G>A, modifies the transcription factor–binding site, whereby the activity of the G allele is over 40% greater compared to the A allele [182] (see Table 9.2). Furthermore, Rieder et al. also found reduced amounts of VKORC1 mRNA in carriers of the A allele [180], resulting in reduced numbers of functional VKORC1 enzyme, the rate-limiting factor in the vitamin K pathway. Another well-known polymorphism is located on the first intron of VKORC1. It is called C6484T, or 1173C>T, and is found in almost perfect linkage disequilibrium with G3673A, meaning the A allele in G3673A and the T allele in

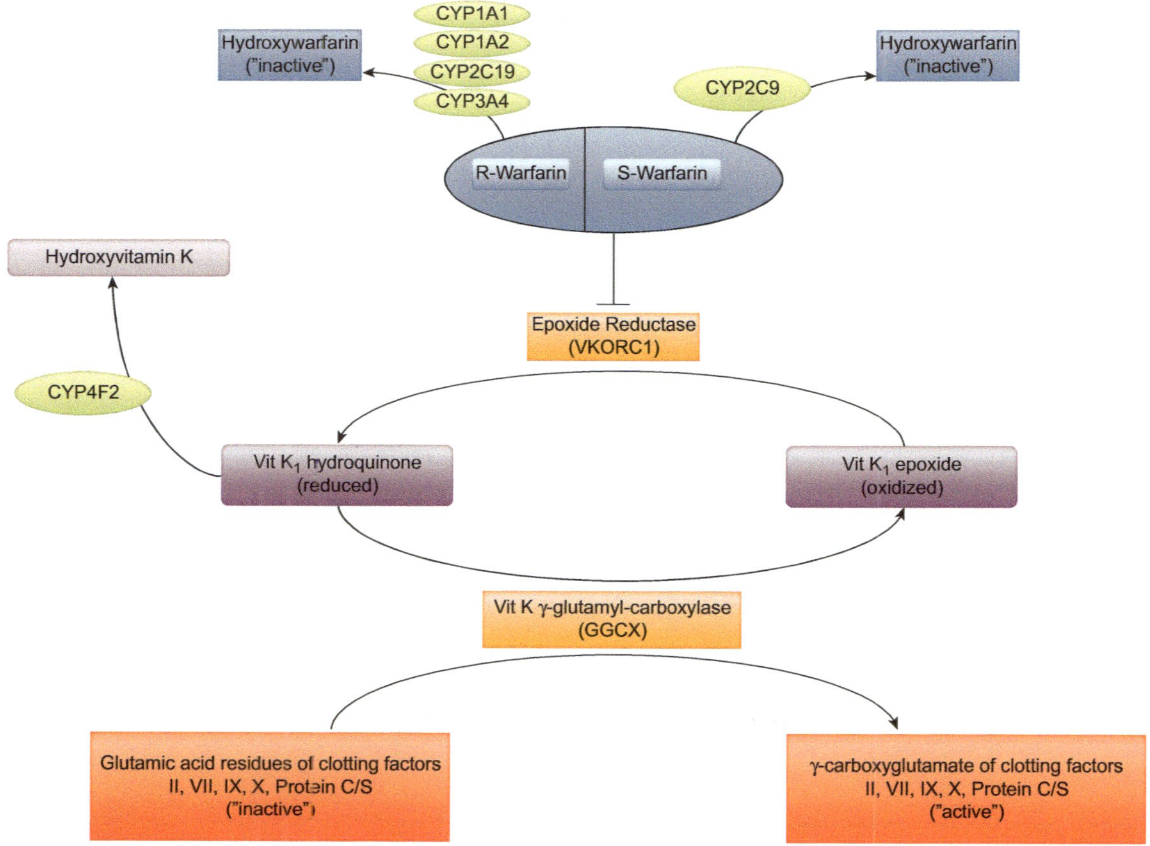

FIGURE 9.4 **Simplified vitamin K cycle, its influence on clotting factors, and the effect of warfarin.** *Recreated using data from sources [164–179].*

TABLE 9.2 Polymorphisms of VKORC1 [7,8,13,14]

Haplotype [8]	SNP	Haplotype Group [7]	VKORC1 Activity	Mean Warfarin Dose	Population
VKORC1*1	Wild-type		Normal	Intermediate	General population
VKORC1*2	G3673A (–1639G>A)	A	Reduced	Low	85–90% Asian 38–43% Caucasian 8–13% African
VKORC1*2	C6484T (1173C>T)	A	Reduced	Low	85–90% Asian 38–43% Caucasian 8–13% African
VKORC1*3	G9041A (3730G>A)	B	Increased	High	10–15% Asian 20–39% Caucasian 42–53% African
VKORC1*4	C6009T	B	Increased	High	<2% Asian 20–38% Caucasian 7–11% African

C6484T are generally inherited together [165]. In fact, these two alleles define the VKORC1*2 haplotype as described by Geisen et al. [181,183], or the two haplotypes VKORC1 H1 and H2 as defined by Rieder et al. [180], classified as the warfarin low-dose haplotype group A. Carriers of this haplotype have been shown in different studies to require a lower mean warfarin dose than carriers of the reference wild-type VKORC1*1, and, even more strikingly, carriers of haplotypes from the high-dose group B, such as VKORC1*3 and VKORC1*4 according to Geisen et al.'s classification [181], or haplotypes H7, H8, and H9 according to Rieder et al. [180]. The single nucleotide polymorphism G9041A, or 3730G>A, is located in the three prime untranslated region (3′UTR) of VKORC1 and is associated with the warfarin high-dose haplotype VKORC1*3.

The described effect of genetic influences on the required warfarin dose seems to be additive, in that carriers of two VKORC1*2 low-dose haplotypes A, labeled A/A, respond to the lowest mean warfarin dose; carriers of two high-dose haplotypes from group B, labeled B/B, require the highest doses; and carriers of two mixed haplotypes A/B require an intermediate mean warfarin dose [180].

Moreover, there seems to not only be an interindividual but also an interethnic variability. The VKORC1*2 haplotype is the predominant haplotype in the Asian population [180,182], who have been known to require a lower warfarin maintenance dose than other ethnicities, even after weight adjustment [184,185]. Haplotype VKORC1*3, in contrast, is most common in people of African ethnicity, shown in studies to require higher warfarin maintenance doses [186].

In addition to the described effects of multiple VKORC1 haplotypes, genetic influences on the metabolism of warfarin directly contribute approximately 12% to interindividual differences in response to warfarin therapy [168]. Warfarin, as it is administered in clinical practice, is a racemic mixture of two isomers [169], with S-warfarin being a much more potent vitamin K antagonist than its R-enantiomer [170] and, under steady state conditions, accounting for roughly two-thirds of warfarin's anticoagulant effect [167]. In addition to this, S- and R-warfarin slightly differ in their metabolism by cytochrome P450. S-warfarin is principally metabolized by CYP2C9, the main CYP2C in human liver, which is also responsible for the metabolic clearance of numerous other drugs, including phenytoin and other convulsants, the antidiabetic drugs glipizide and tolbutamide, and nonsteroidal anti-inflammatory drugs [171–173]. R-warfarin, on the other hand, is metabolized mainly by CYP3A4, CYP1A1, and CYP1A2 [167]. Currently, most of the data is available for CYP2C9, particularly its two most common allelic variants CYP2C9*2, or 430C>T, and CYP2C9*3, also labeled 1075A>C, although dozens of different CYP2C9 variants have been identified to this date [187]. Both CYP2C9*2 and CYP2C9*3 feature a reduced metabolic activity compared to CYP2C9*1, which is considered the wild type and is the most prevalent CYP2C9 allele [188]. The maximum metabolism rate of CYP2C9*2, for example, is only about half of the CYP2C9*1 rate [167]. Carriers of a reduction-of-function variant, either CYP2C9*2 or CYP2C9*3, therefore require lower warfarin doses than carriers of the wild type, and have a higher risk of bleeding complications, which were most obvious during the initiation and dose titration of warfarin therapy [189,190].

Later studies suggested that, in addition to the influences of VKORC1 and CYP2C9, about 1–2% of warfarin dose variance might be attributed to polymorphisms in cytochrome P450 4F2 (CYP4F2), affecting a different pathway in the vitamin K cycle [168]. CYP4F2 was shown to be involved in vitamin K catabolism by oxidizing vitamin K_1 to hydroxyvitamin K_1. The V433M polymorphism (rs2108622) of the CYP4F2 gene is thought to be associated with limited translation or degradation of CYP4F2, while seemingly leaving the enzyme's intrinsic catalytic activity unaffected [174]. Carriers of this variant have a reduced ability to catabolize vitamin K_1, resulting in raised hepatic levels of vitamin K and a warfarin dose requirement about 1 mg/day higher than that of individuals carrying the CYP4F2 wild type [175].

In recent years, a number of studies have examined the effects of polymorphisms in the GGCX gene, encoding the vitamin-K-dependent gamma-glutamyl carboxylase on warfarin dose requirements and dose variance. The enzyme, found in the membrane of rough endoplasmic reticulum, carboxylates glutamic acid residues of vitamin K-dependent proteins, such as the clotting factors II, VII, IX, and X, to calcium-binding gamma-carboxyglutamate residues, an important step required for their activation [176]. Different studies focused on different gene variants and different ethnic groups. Cavallari et al. [177], for example, suggested that the GGCX genotype rs10654848, characterized by a (CAA)16/17 repeat polymorphism, is 10-fold more common in African Americans than it is in Caucasians, and was more frequent in individuals requiring warfarin doses of 7.5 mg/day or greater. Another study by Huang et al. [178] focused on the GGCX 3261G>A polymorphism and concluded that, among a group of Chinese patients on stable warfarin treatment, carriers of the 3261AA genotype had a significantly higher daily warfarin dose requirement than individuals with the 3261GG genotype. A systematic review and meta-analysis from 2015 by Sun et al. [179] summarized that, while certain GGCX polymorphisms have indeed been shown to influence warfarin dose requirements, this effect seems to differ between ethnicities.

Ever since the discovery of the first genetic influences, apart from age, sex, and weight, on interindividual warfarin dose variability, numerous algorithms have been created, aiming to better predict required warfarin starting and maintenance doses for patients, in order to maximize its therapeutic effect while keeping the risk of bleeding, the most feared adverse effect of warfarin therapy, as low as possible. Some of the most recently developed algorithms combine genetic polymorphisms of VKORC1, CYP2C9, CYP4F2, and GGCX [191]. The majority of studies comparing different algorithms have shown that those incorporating both clinical information and genetic data have a better predictive ability compared with those using clinical parameters alone [192]. In spite of this, clinical applicability of these more complex dosing algorithms has not yet been established. One reason is that some studies relating to pharmacogenetics-guided dosing regimens have, in fact, not shown any improved outcomes [193]. Another factor making pharmacogenomics-based warfarin dose algorithms somewhat less feasible is its arguable cost effectiveness, and future studies will certainly give further information regarding the practicability of pharmacogenetics-guided algorithms in an everyday clinical setting.

REFERENCES

[1] Maron BJ, Gardin JM, Flack JM, Gidding SS, Kurosaki TT, Bild DE. Prevalence of hypertrophic cardiomyopathy in a general population of young adults. Echocardiographic analysis of 4111 subjects in the CARDIA Study. Coronary Artery Risk Development in (Young) Adults. Circulation 1995;92:785–9.

[2] Maron BJ. Hypertrophic cardiomyopathy: a systematic review. JAMA 2002;287:1308–20.

[3] Maron BJ, McKenna WJ, Danielson GK, Kappenberger LJ, Kuhn HJ, Seidman CE, et al. American College of Cardiology/European Society of Cardiology clinical expert consensus document on hypertrophic cardiomyopathy. A report of the American College of Cardiology task force on clinical expert consensus documents and the European Society of Cardiology Committee for practice guidelines committee to develop an expert consensus document on hypertrophic cardiomyopathy. J Am Coll Cardiol 2003;42:1687–713.

[4] Maron BJ. Hypertrophic cardiomyopathy: an important global disease. Am J Med 2004;116: 63–5.

[5] Teare D. Asymmetrical hypertrophy of the heart in young adults. Br Heart J 1958;20:1–8.

[6] Maron BJ, Seidman CE, Ackerman MJ, Towbin JA, Maron MS, Ommen SR, et al. How should hypertrophic cardiomyopathy be classified?: what's in a name? Dilemmas in nomenclature characterizing hypertrophic cardiomyopathy and left ventricular hypertrophy. Circ Cardiovasc Genet 2009;2:81–5.

[7] Maron BJ, Towbin JA, Thiene G, Antzelevitch C, Corrado D, Arnett D, et al. Contemporary definitions and classification of the cardiomyopathies: an American heart association scientific statement from the council on clinical cardiology, heart failure and transplantation committee; quality of care and outcomes research and functional genomics and translational biology interdisciplinary working groups; and council on epidemiology and prevention. Circulation 2006;113:1807–16.

[8] Maron BJ, Doerer JJ, Haas TS, Tierney DM, Mueller FO. Sudden deaths in young competitive athletes: analysis of 1866 deaths in the United States, 1980–2006. Circulation 2009;119:1085–92.

[9] Gersh BJ, Maron BJ, Bonow RO, Dearani JA, Fifer MA, Link MS, et al. 2011 ACCF/AHA guideline for the diagnosis and treatment of hypertrophic cardiomyopathy: executive summary: a report of the American College of Cardiology Foundation/America heart association task force on practice guidelines. Circulation 2011;124:2761–96.

[10] Alcalai R, Seidman JG, Seidman CE. Genetic basis of hypertrophic cardiomyopathy: from bench to the clinics. J Cardiovasc Electrophysiol 2008;19:104–10.

[11] Van Driest SL, Ommen SR, Tajik AJ, Gersh BJ, Ackerman MJ. Yield of genetic testing in hypertrophic cardiomyopathy. Mayo Clin Proc 2005;80:739–44.

[12] Sherrid MV. Pathophysiology and treatment of hypertrophic cardiomyopathy. Prog Cardiovasc Dis 2006;49:123–51.

[13] Cirino AL, Ho C. Hypertrophic cardiomyopathy overview. [Updated January 16, 2014] [Internet] In: Pagon RA, Adam MP, Ardinger HH, et al., editors. GeneReviews®. Seattle (WA): University of Washington; August 5, 2008. Seattle; 1993–2015. Available from: http://www.ncbi.nlm.nih.gov/books/NBK1768/.

[14] Tajsharghi H. Thick and thin filament gene mutations in striated muscle diseases. Int J Mol Sci 2008;9:1259–75.

[15] Cooper GM. . The cell: A molecular approach. 2nd ed. Sunderland (MA): Sinauer Associates; 2000. Actin, Myosin, and Cell Movement. Available from: http://www.ncbi.nlm.nih.gov/books/NBK9961/. [chapter 15].

[16] Noureldin RA, Liu S, Nacif MS, Judge DP, Halushka MK, Abraham TP, et al. The diagnosis of hypertrophic cardiomyopathy by cardiovascular magnetic resonance. J Cardiovasc Magn Reson 2012;14:17.

[17] Cecchi F, Olivotto I, Gistri R, Lorenzoni R, Chiriatti G, Camici PG. Coronary microvascular dysfunction and prognosis in hypertrophic cardiomyopathy. N Engl J Med 2003;349:1027–35.

[18] Maron MS, Olivotto I, Betocchi S, Casey SA, Lesser J, Losi MA, et al. Effect of left ventricular outflow tract obstruction on clinical outcome in hypertrophic cardiomyopathy. N Engl J Med 2003;348:295–303.

[19] Walker CM, Reddy GP, Mohammed TL, Chung JH. Systolic anterior motion of the mitral valve. J Thorac Imaging 2012;27:W87.

[20] Poliac LC, Barron ME, Maron BJ. Hypertrophic cardiomyopathy. Anesthesiology 2006;104:183–92.

[21] Schwartz PJ, Ackerman MJ, George AL, Wilde AA. Impact of genetics on the clinical management of channelopathies. J Am Coll Cardiol 2013;62:169–80.

[22] Tester DJ, Ackerman MJ. Postmortem long QT syndrome genetic testing for sudden unexplained death in the young. J Am Coll Cardiol 2007;49:240–6.

[23] Arnestad M, Crotti L, Rognum TO, Insolia R, Pedrazzini M, Ferrandi C, et al. Prevalence of long QT syndrome gene variants in sudden infant death syndrome. Circulation 2007;115:361–7.

[24] Guidicessi JR, Ackerman MJ. Prevalence and potential genetic determinants of sensorineural deafness in KCNQ1 homozygosity and compound heterozygosity. Circ Cardiovasc Genet 2013;6:193–200.

[25] Keating MT, Atkinson D, Dunn C, Timothy K, Vincent GM, Leppert M. Consistant linkage of the long QT syndrome to the Harvey ras-1 locus on chromosome 11. Am J Hum Genet 1991;49:1335–9.

[26] Jiang C, Atkinson D, Towbin JA, Splawski I, Lehmann MH, Li H, et al. Two long QT syndrome loci map to chromosomes 3 and 7 with evidence for further heterogeneity. Nature Genet 1994;6:141–7.

[27] Ackerman MJ. Genotype-phenotype relationships in congenital long QT syndrome. J Electrocardiol 2005;38:64–8.

[28] Grant AO. Cardiac ion channels. Circ Arrhythm Electrophysiol 2009;2:185–94.

[29] Ravens U, Wettwer E. Ultra-rapid delayed rectifier channels: molecular basis and therapeutic implications. Cardiovasc Res 2011;89:776–85.

[30] Gutman GA, Chandy KG, Grissmer S, Lazdunski M, McKinnon D, Pardo LA, et al. International Union of Pharmacology. LIII. Nomenclature and molecular relationships of voltage-gated potassium channels. Pharmacol Rev 2005;57:473–508.

[31] Bennett PB, Yazawa K, Makita N, George Jr AL. Molecular mechanism for an inherited cardiac arrhythmia. Nature 1995;376:683–5.

[32] Tester DJ, Will ML, Haglund CM, Ackerman MJ. Compendium of cardiac channel mutations in 541 consecutive unrelated patients referred for long QT syndrome genetic testing. Heart Rhythm 2005;2:507–17.
[33] Tester DJ, Ackerman MJ. Genetics of long QT syndrome. Methodist Debakey Cardiovasc J 2014;10:29–33.
[34] Splawski I, Tristani-Firouzi M, Lehmann MH, Sanguinetti MC, Keating MT. Mutations in the hminK gene cause long QT syndrome and suppress I_{Ks} function. Nat Genet 1997;17:338–40.
[35] Abbott GW, Sesti F, Splawski I, Buck ME, Lehmann MH, Timothy KW, et al. MiRP1 forms I_{Kr} potassium channels with HERG and is associated with cardiac arrhythmia. Cell 1999;97:175–87.
[36] Medeiros-Domingo A, Kaku T, Tester DJ, Iturralde-Torres P, Itty A, Ye B, et al. SCN4B-encoded sodium channel beta4 subunit in congenital long-QT syndrome. Circulation 2007;116:134–42.
[37] Yang Y, Yang Y, Liang B, Liu J, Li J, Grunnet M, et al. Identification of a Kir3.4 mutation in congenital long QT syndrome. Am J Hum Genet 2010;86:872–80.
[38] Beckmann C, Rinne A, Littwitz C, Mintert E, Bosche LI, Kienitz MC, et al. G protein-activated (GIRK) current in rat ventricular myocytes is masked by constitutive inward rectifier current (I(K1)). Cell Physiol Biochem 2008;21:259–68.
[39] Schott JJ, Charpentier F, Peltier S, Foley P, Drouin E, Bouhour JB, et al. Mapping of a gene for long QT syndrome to chromosome 4q25-27. Am J Hum Genet 1995;57:1114–22.
[40] Curran J, Mohler PJ. Coordinating electrical activity of the heart: ankyrin polypeptides in human cardiac disease. Expert Opin Ther Targets 2011;15:789–801.
[41] Bennett V, Baines AJ. Spectrin and ankyrin-based pathways: metazoan inventions for integrating cells into tissues. Physiol Rev 2001;81:1353–92.
[42] Mohler PJ, Le Scouarnec S, Denjoy I, Lowe JS, Guicheney P, Caron L, et al. Defining the cellular phenotype of "ankyrin-B syndrome" variants: human ANK2 variants associated with clinical phenotypes display a spectrum of activities in cardiomyocytes. Circulation 2007;115:432–41.
[43] Maltsev VA, Lakatta EG. Normal heart rhythm is initiated and regulated by an intracellular calcium clock within pacemaker cells. Heart Lung Circ 2007;16:335–48.
[44] Healy JA, Nilsson KR, Hohmeier HE, Berglund J, Davis J, Hoffman J, et al. Cholinergic augmentation of insulin release requires ankyrin-B. Sci Signal 2010;3:ra19.
[45] Hund TJ, Wright PJ, Dun W, Snyder JS, Boyden PA, Mohler PJ. Regulation of the ankyrin-B-based targeting pathway following myocardial infarction. Cardiovasc Res 2009;81:742–9.
[46] Hedley PL, Kanters JK, Dembic M, Jespersen T, Skibsbye L, Aidt FH, et al. The role of CAV3 in long-QT syndrome: clinical and functional assessment of a caveolin-3/K_V11.1 double heterozygote versus caveolin-3 single heterozygote. Circ Cardiovasc Genet 2013;6:452–61.
[47] Gazzerro E, Sotgia F, Bruno C, Lisanti MP, Minetti C. Caveolinopathies: from the biology of caveolin-3 to human diseases. Eur J Hum Genet 2010;18:137–45.
[48] Vaidyanathan R, Vega AL, Song C, Zhou Q, Tan B, Berger S, et al. The interaction of caveolin 3 protein with the potassium inward rectifier channel Kir2.1: physiology and pathology related to long QT syndrome 9 (LQT9). J Biol Chem 2013;288:17472–80.
[49] Li Y, Chen L, Kass RS, Dessauer CW. The A-kinase anchoring protein Yotiao facilitates complex formation between adenylyl cyclase type 9 and the I_{Ks} potassium channel in heart. J Biol Chem 2012;287:29815–24.
[50] Chen L, Marquardt ML, Tester DJ, Sampson KJ, Ackerman MJ, Kass RS. Mutation of an A-kinase-anchoring protein causes long-QT syndrome. Proc Natl Acad Sci U S A 2007;104:20990–5.

[51] Ahn AH, Freener CA, Gussoni E, Yoshida M, Ozawa E, Kunkel LM. The three human syntrophin genes are expressed in diverse tissues, have distinct chromosomal locations, and each bind to dystrophin and its relatives. J Biol Chem 1996;271:2724–30.

[52] Williams JC, Armesilla AL, Mohamed TM, Hagarty CL, McIntyre FH, Schomburg S, et al. The sarcolemmal calcium pump, alpha-1 syntrophin, and neuronal nitric-oxide synthase are parts of a macromolecular protein complex. J Biol Chem 2006;281:23341–8.

[53] Ueda K, Valdivia C, Medeiros-Domingo A, Tester DJ, Vatta M, Farrugia G, et al. Syntrophin mutation associated with long QT syndrome through activation of the nNOS-SCN5A macromolecular complex. Proc Natl Acad Sci U S A 2008;105:9355–60.

[54] Hu R-M, Tan B-H, Orland KM, Valdivia CR, Peterson A, Pu J, et al. Digenic inheritance novel mutations in SCN5a and SNTA1 increase late I_{Na} contributing to LQT syndrome. Am J Physiology Heart Circ Physiol 2013;304:H994–1001.

[55] Zhang M, D'Aniello C, Verkerk AO, Wrobel E, Frank S, Ward-van Oostwaard D, et al. Recessive cardiac phenotypes in induced pluripotent stem cell models of Jervell and Lange-Nielsen syndrome: disease mechanisms and pharmacological rescue. Proc Natl Acad Sci 2014;111:E5383–92.

[56] Schwartz PJ, Spazzolini C, Crotti L, Bathen J, Amlie JP, Timothy K, et al. The Jervell and Lange-Nielsen syndrome: natural history, molecular basis, and clinical outcome. Circulation 2006;113:783–90.

[57] Chang R-KR, Lan Y-T, Silka MJ, Morrow H, Kwong A, Smith-Lang J, et al. Genetic variants for long QT syndrome among infants and children from a statewide newborn hearing screening program cohort. J Pediat 2014;164:590–5.

[58] Wrobel E, Tapken D, Seebohm G. The KCNE Tango – how KCNE1 interacts with K_V7.1. Front Pharmacol 2012;3:142.

[59] Yoon G, Oberoi S, Tristani-Firouzi M, Etheridge SP, Quitania L, Kramer JH, et al. Andersen-Tawil syndrome: prospective cohort analysis and expansion of the phenotype. Am J Med Genet A 2006;140:312–21.

[60] Bendahhou S, Donaldson MR, Plaster NM, Tristani-Firouzi M, Fu YH, Ptacek LJ. Defective potassium channel Kir2.1 trafficking underlies Andersen-Tawil syndrome. J Biol Chem 2003;278:51779–85.

[61] Hasegawa K, Ohno S, Kimura H, Itoh H, Makiyama T, Yoshida Y, et al. Mosaic KCNJ2 mutation in Andersen-Tawil syndrome: targeted deep sequencing is useful for the detection of mosaicism. Clin Genet 2015;87:279–83.

[62] Tristani-Firouzi M, Jensen JL, Donaldson MR, Sansone V, Meola G, Hahn A, et al. Functional and clinical characterization of KCNJ2 mutations associated with LQT7 (Andersen syndrome). J Clin Invest 2002;110:381–8.

[63] Zhang L, Benson DW, Tristani-Firouzi M, Ptacek LJ, Tawil R, Schwartz PJ, et al. Electrocardiographic features in Anderson-Tawil syndrome patients with KCNJ2 mutations: characteristic T-U-wave patterns predict the KCNJ2 genotype. Circulation 2005;111:2720–6.

[64] Schwartz PJ, Crotti L, Insolia R. Long QT syndrome: from genetics to management. Circ Arrhythm Electrophysiol 2012;5:868–77.

[65] Splawski I, Timothy KW, Sharpe LM, Decher N, Kumar P, Bloise R, et al. Ca(V)1.2 calcium channel dysfunction causes a multisystem disorder including arrhythmia and autism. Cell 2004;119:19–31.

[66] Splawski I, Timothy KW, Decher N, Kumar P, Sachse FB, Beggs AH, et al. Severe arrhythmia disorder caused by cardiac L-type calcium channel mutations. Proc Natl Acad Sci U S A 2005;102:8089–96.

[67] Ertel EA, Campbell KP, Harpold MM, Hofmann F, Mori Y, Perez-Reyes E, et al. Nomenclature of voltage-gated calcium channels. Neuron 2000;25:533–5.

[68] Hofmann F, Flockerzi V, Kahl S, Wegener JW. L-type CaV1.2 calcium channels: from in vitro findings to in vivo function. Physiol Rev 2014;94:303–26.

[69] Barrett CF, Tsien RW. The Timothy syndrome mutations differentially affects voltage- and calcium-dependent inactivation of CaV1.2 L-type calcium channels. Proc Natl Acad Sci U S A 2008;105:2157–62.

[70] George AL. Molecular and genetic basis of sudden cardiac death. J Clin Invest 2013;123: 75–83.

[71] Liu N, Ruan Y, Priori SG. Catecholaminergic polymorphic ventricular tachycardia. Prog Cardiovasc Dis 2008;51:23–30.

[72] Tester DJ, Kopplin LJ, Will ML, Ackerman MJ. Spectrum and prevalence of cardiac ryanodine receptor (RyR2) mutations in a cohort of unrelated patients referred explicitly for long QT syndrome genetic testing. Heart Rhythm 2005;2:1099–105.

[73] Horner JM, Ackerman MJ. Ventricular ectopy during treadmill exercise stress testing in the evaluation of long QT syndrome. Heart Rhythm 2008;5:1690–4.

[74] Tester DJ, Dura M, Carturan E, Reiken S, Wronska A, Marks AR, et al. A mechanism for sudden infant death syndrome (SIDS): stress-induced leak via ryanodine receptors. Heart Rhythm 2007;4:733–9.

[75] Stokes DL, Wagenknecht T. Calcium transport across the sarcoplasmic reticulum: structure and function of Ca^{2+}-ATPase and the ryanodine receptor. Eur J Biochem 2000;267:5274–9.

[76] Meissner G. Regulation of mammalian ryanodine receptors. Front Biosci 2002;7:d2072–80.

[77] Masumiya H, Wang R, Zhang J, Xiao B, Chen SR. Localization of the 12.6-kDa FK506-binding protein (FKBP12.6) binding site to the NH2-terminal domain of the cardiac Ca^{2+} release channel (ryanodine receptor). J Biol Chem 2003;278:3786–92.

[78] Marx SO, Reiken S, Hisamatsu Y, Jayaraman T, Burkhoff D, Rosemblit N, et al. PKA phosphorylation dissociates FKBP12.6 from the calcium release channel (ryanodine receptor): defective regulation in failing hearts. Cell 2000;101:365–76.

[79] Tung CC, Lobo PA, Kimlicka L, Van Petegem F. The amio-terminal disease hotspot of ryanodine receptors forms a cytoplasmic vestibule. Nature 2010;468:585–8.

[80] Wehrens XH, Lehnart SE, Huang F, Vest JA, Reiken SR, Mohler PJ, et al. FKBP12.6 deficiency and defective calcium release channel (ryanodine receptor) function linked to exercise-induced sudden cardiac death. Cell 2003;113:829–40.

[81] George CH, Higgs GV, Lai FA. Ryanodine receptor mutations associated with stress-induced ventricular tachycardia mediate increased calcium release in stimulated cardiomyocytes. Circ Res 2003;93:531–40.

[82] Jiang D, Wang R, Xiao B, Kong H, Hunt DJ, Choi P, et al. Enhanced store overload-induced Ca^{2+} release and channel sensitivity to luminal Ca^{2+} activation are common defects of RyR2 mutations linked to ventricular tachycardia and sudden death. Circ Res 2005;97:1173–81.

[83] Faggioni M, Kryshtal DO, Knollmann BC. Calsequestrin mutations and catecholaminergic polymorphic ventricular tachycardia. Pediatr Cardiol 2012;33:959–67.

[84] Viatchenko-Karpinski S, Terentyev D, Gyorke I, Terentyeva R, Volpe P, Priori SG, et al. Abnormal calcium signaling and sudden cardiac death associated with mutation of calsequestrin. Circ Res 2004;94:471–7.

[85] Brugada P, Brugada J. Right bundle branch block, Persistent ST segment elevation and sudden cardiac death: a distinct clinical and electrocardiographic syndrome. A multicenter report. J Am Coll Cardiol 1992;20:1391–6.

[86] Atarashi H, Ogawa S, Harumi K, Sugimoto T, Inoue H, Murayama M, et al. Three-year follow-up of patients with right bundle branch block and ST segment elevation in the right precordial leads: Japanese Registry of Brugada syndrome. Idiopathic Ventricular Fibrillation Investigators. J Am Coll Cardiol 2001;37:1916–20.
[87] Hermida JS, Lemoine JL, Aoun FB, Jarry G, Rey JL, Quiret JC. Prevalence of the Brugada syndrome in an apparently healthy population. Am J Cardiol 2000;86:91–4.
[88] Atarashi H, Ogawa S, Harumi K, Hayakawa H, Sugimoto T, Okada R, et al. Characteristics of patients with right bundle branch block and ST-segment elevation in right precordial leads. Idiopathic Ventricular Fibrillation Investigators. Am J Cardiol 1996;78:581–3.
[89] Wilde AA, Antzelevitch C, Borggrefe M, Brugada J, Brugada R, Brugada P, et al. Proposed diagnostic criteria for the Brugada syndrome: consensus report. Circulation 2002;106:2514–9.
[90] Meregalli PG, Wilde AA, Tan HL. Pathophysiological mechanisms of Brugada syndrome: depolarization disorder, repolarization disorder, or more? Cardiovasc Res 2005;67:367–78.
[91] Kapplinger JD, Tester DJ, Alders M, Benito B, Berthet M, Brugada J, et al. An international compendium of mutations in the SCN5A-encoded cardiac sodium channel in patients referred for Brugada syndrome genetic testing. Heart Rhythm 2010;7:33–46.
[92] Boussy T, Paparella G, De Asmundis C, Sarkozy A, Chierchia GB, Brugada J, et al. Genetic basis of ventricular arrhythmias. Cardiol Clin 2008;26:335–53.
[93] Berne P, Brugada J. Brugada syndrome 2012. Circ J 2012;76:1563–71.
[94] Bezzina C, Veldkamp MW, van den Berg MP, Postma AV, Rook MB, Viersma JW, et al. A single Na(+) channel mutation causing both long-QT and Brugada syndromes. Circ Res 1999;85:1206–13.
[95] Antzelevitch C, Pollevick GD, Cordeiro JM, Casis O, Sanguinetti MC, Aizawa Y, et al. Loss-of-function mutations in the cardiac calcium channel underlie a new clinical entity characterized by ST-segment elevation, short QT intervals, and sudden cardiac death. Circulation 2007;115:442–9.
[96] Brugada P. Commentary on the Brugada ECG pattern: a marker of channelopathy, structural heart disease, or neither? toward a unifying mechanism of the Brugada syndrome. Circ Arrhythm Electrophysiol 2010;3:280–2.
[97] Yan GX, Antzelevitch C. Cellular basis for the Brugada syndrome and other mechanisms of arrhythmogenesis associated with ST-segment elevation. Circulation 1999;100:1660–6.
[98] Kumar A, Agarwal S. Marfan syndrome: an eyesight of syndrome. Meta Gene 2014;2:96–105.
[99] Judge DP, Dietz HC. Marfan's syndrome. Lancet 2005;366:1965–76.
[100] Keane MG, Pyeritz RE. Medical management of Marfan syndrome. Circulation 2008;117:2802–13.
[101] Pereira L, Lee SY, Gayraud B, Andrikopoulos K, Shapiro SD, Bunton T, et al. Pathogenetic sequence for aneurysm revealed in mice underexpressing fibrillin-1. Proc Natl Acad Sci U S A 1999;96:3819–23.
[102] El-Hamamsy I, Yacoub MH. Cellular and molecular mechanisms of thoracic aortic aneurysms. Nat Rev Cardiol 2009;6:771–86.
[103] Bunton TE, Biery NJ, Myers L, Gayraud B, Ramirez F, Dietz HC. Phenotypic alteration of vascular smooth muscle cells precedes elastolysis in a mouse model of Marfan syndrome. Circ Res 2001;88:37–43.
[104] Massagué J. The TGF-beta family of growth and differentiation factors. Cell 1987;49:437–8.
[105] Kaartinen V, Warburton D. Fibrillin controls TGF-beta activation. Nat Genet 2003;33:331–2.
[106] Dyhdalo K, Farver C. Pulmonary histologic changes in Marfan syndrome: a case series and literature review. Am J Clin Pathol 2011;136:857–63.

[107] Neptune ER, Frischmeyer PA, Arking DE, Myers L, Bunton TE, Gayraud B, et al. Dysregulation of TGF-beta activation contributes to pathogenesis in Marfan syndrome. Nat Genet 2003;33:407–11.

[108] Ng CM, Cheng A, Myers LA, Martinez-Murillo F, Jie C, Bedja D, et al. TGF-beta dependent pathogenesis of mitral valve prolapse in a mouse model of Marfan syndrome. J Clin Invest 2004;114:1586–92.

[109] Loeys BL, Chen J, Neptune ER, Judge DP, Podowski M, Holm T, et al. A syndrome of altered cardiovascular, craniofacial, neurocognitive and skeletal development caused by mutations in TGFBR1 or TGFBR2. Nat Genet 2005;37:275–81.

[110] MacCarrick G, Black JH, Bowdin S, El-Hamamsy IE, Frischmeyer-Guerrerio PA, Guerrerio AL, et al. Loeys–Dietz syndrome: a primer for diagnosis and management. Genet Med 2014;16:576–87.

[111] Disabella E, Grasso M, Marziliano N, Ansaldi S, Lucchelli C, Porcu E, et al. Two novel and one known mutation of the TGFBR2 gene in Marfan syndrome not associated with FBN1 gene defects. Eur J Hum Genet 2006;14:34–8.

[112] Moustakas A, Souchelnytskyi S, Heldin CH. Smad regulation in TGF-beta signal transduction. J Cell Sci 2001;114:4359–69.

[113] Lindsay ME, Schepers D, Ajit Bolar N, Doyle J, Gallo E, Fert-Bober J, et al. Loss-of-function mutations in TGFB2 cause a syndromic presentation of thoracic aortic aneurysm. Nat Genet 2012;44:922–7.

[114] Gallo EM, Loch DC, Habashi JP, Calderon JF, Chen Y, Bedja D, et al. Angiotensin II-dependent TGF-β signaling contributes to Loeys-Dietz syndrome vascular pathogenesis. J Clin Invest 2014;124:448–60.

[115] Lusis AJ, Fogelman AM, Fonarow GC. Genetic basis of atherosclerosis: Part I. Circulation 2004;110:1868–73.

[116] The Nobel Prize in Physiology or Medicine 1985. Nobelprize.org. Nobel Media AB 2014 http://www.nobelprize.org/nobel_prizes/medicine/laureates/1985/ [web 11.11.15].

[117] Brown MS, Goldstein JL. Familial hypercholesterolemia: defective binding of lipoproteins to cultured fibroblasts associated with impaired regulation of 3-hydroxy-3-methylglutary coenzyme A reductase activity. Proc Natl Acad Sci U S A 1974;71:788–92.

[118] Hovingh GK, Davidson MH, Kastelein JJ, O'Connor AM. Diagnosis and treatment of familial hypercholesterolaemia. Eur Heart J 2013;34:962–71.

[119] Hopkins PN, Toth PP, Ballantyne CM, Rader DJ. Familial hypercholesterolemias: prevalence, genetics, diagnosis and screening recommendations from the National Lipid Association Expert Panel on familial hypercholesterolemia. J Clin Lipidol 2011;5:9–17.

[120] Kovacic S, Bakran M. Genetic susceptibility to atherosclerosis. Stroke Res Treat 2012;2012:362941.

[121] Bourbon M, Duarte MA, Alves AC, Medeiros AM, Marques L, Soutar AK. Genetic diagnosis of familial hypercholesterolaemia: the importance of functional analysis of potential splice-site mutations. J Med Genet 2009;46:352–7.

[122] Sijbrands EJ, Westendorp RG, Defesche JC, de Meier PH, Smelt AH, Kastelein JJ. Mortality over two centuries in large pedigree with familial hypercholesterolaemia: family tree mortality study. BMJ 2001;322:1019–23.

[123] Goldstein JL, Hobbs HH, Brown MS. Familial hypercholesterolaemia. In: 8th ed. Scriver CR, Beaudet AL, Sly WS, Vale D, editors. The metabolic and molecular basis of inherited disease, vol. III. New York: McGraw Hill; 2001. p. 2863e914.

[124] Stock J. New EAS consensus statement on FH: improving the care of FH patients. Atherosclerosis 2013;231:69–71.

[125] Lusis A. Atherosclerosis. Nature 2000;407:233–41.

[126] Mehrabian M, Allayee H. 5-lipoxygenase and atherosclerosis. Curr Opin Lipidol 2003;14:447–57.

[127] Funk CD, Chen XS. 5-lipoxygenase and leukotrienes: transgenic mouse and nuclear targeting studies. Am J Respir Crit Care Med 2000;161:S120–4.

[128] Nigam S, Pace-Asciak FR. Lipoxygenases and their metabolites: biological functions. New York, NY: Plenum; 1999.

[129] Hofker M. Lipoxygenases at the heart of atherosclerosis susceptibility. Eur J Hum Genet 2004;12:601–2.

[130] Mehrabian M, Wong J, Wang X, Jiang Z, Shi W, Fogelman A, et al. Genetic locus in mice that blocks development of atherosclerosis despite extreme hyperlipidemia. Circ Res 2001;89:125–30.

[131] Mehrabian M, Allayee H, Wong J, Shi W, Wang XP, Shaposhnik Z, et al. Identification of 5- lipoxygenase as a major gene contributing to atherosclerosis susceptibility in mice. Circ Res 2002;91:120–6.

[132] Dwyer JH, Allayee H, Dwyer KM, Fan J, Wu H, Mar R, et al. Arachidonate 5-lipoxygenase promoter genotype, dietary arachidonic acid, and atherosclerosis. N Engl J Med 2004;350:29–37.

[133] Helgadottir A, Manolescu A, Thorleifsson G, Gretarsdottir S, Jonsdottir H, Thorsteinsdottir U, et al. The gene encoding 5-lipoxygenase activating protein confers risk of myocardial infarction and stroke. Nat Genet 2004;36:233–9.

[134] Wang L, Fan C, Topol SE, Topol EJ, Wang Q. Mutation of MEF2A in an inherited disorder with features of coronary artery disease. Science 2003;302:1578–81.

[135] Edmondson DG, Lyons GE, Martin JF, Olson EN. Mef2 gene expression marks the cardiac and skeletal muscle lineages during mouse embryogenesis. Development 1994;120:1251–63.

[136] Bhagavatula MRK, Fan C, Shen GQ, Cassano J, Plow EF, Topol EJ, et al. Transcription factor MEF2A mutations in patients with coronary artery disease. Hum Mol Genet 2004;13:3181–8.

[137] Weng L, Kavaslar N, Ustaszewska A, Doelle H, Schackwitz W, Hebert S, et al. Lack of MEF2Amutations in coronary artery disease. J Clin Invest 2005;115:1016–20.

[138] Watkins H, Farrall M. Genetic susceptibility to coronary artery disease: from promise to progress. Nat Rev Genet 2006;7:163–73.

[139] Fredrickson DS, Altrocchi PH, Avioli LV, Goodman DS, Goodman HC. Tangier disease: combined clinical staff conference at the National Institute of Health. Ann Intern Med 1961;55:1016–122.

[140] Chiu LS. When a gene makes you smell like a fish…and other tales about the genes in your body. Oxford University Press; 2007.

[141] Rust S, Walter M, Funke H, von Eckardstein A, Cullen P, Kroes HY, et al. Assignment of Tangier disease to chromosome 9q31 by a graphical linkage exclusion strategy. Nat Genet 1998;20:96–8.

[142] Luciani MF, Denizot F, Savary S, Mattei MG, Chimini G. Cloning of two novel ABC transporters mapping on human chromosome 9. Genomics 1994;21:150–9.

[143] Lawn RM, Wade DP, Garvin MR, Wang X, Schwartz K, Porter JG, et al. The Tangier disease gene product ABC1 controls the cellular apolipoprotein-mediated lipid removal pathway. J Clin Invest 1999;104:R25–31.

[144] Miller M, Kwiterovich Jr PO. Isolated low HDL-cholesterol as an important risk factor for coronary heart disease. Eur Heart J 1990;11(Suppl. H):9–14.

[145] Soumian S, Albrecht C, Davies AH, Gibbs RG. ABCA1 and atherosclerosis. Vasc Med 2005;10:109–19.

[146] Puntoni M, Sbrana F, Bigazzi F, Sampietro T. Tangier disease: epidemiology, pathophysiology, and management. Am J Cardiovasc Drugs 2012;12:303–11.

[147] Assman G, von Eckardstein A, Brewer HB. Familial high density lipoprotein deficiency: Tangier disease. In: Scriver CS, Beaudet AL, Sly WS, editors. The metabolic and molecular bases of inherited disease. New York: Mc Graw-Hill; 2001. p. 2053–72.

[148] Sampietro T, Puntoni M, Bigazzi F, Pennato B, Sbrana F, Dal Pino B, et al. Tangier disease in severely progressive coronary and peripheral artery disease. Circulation 2009;119: 2741–2.

[149] Clee SM, Kastelein JJP, van Dam M, Marcil M, Roomp K, Zwarts KY, et al. Age and residual cholesterol efflux affect HDL cholesterol levels and coronary artery disease in ABCA1 heterozygotes. J Clin Invest 2000;106:1263–70.

[150] Navab M, Reddy ST, Van Lenten BJ, Fogelman AM. HDL and cardiovascular disease: atherogenic and atheroprotective mechanisms. Nat Rev Cardiol 2001;8:222–32.

[151] Joy T, Hegele RA. Is raising HDL a futile strategy for atheroprotection? Nat Rev Drug Discov 2008;7:145–55.

[152] Bhattacharyya AK, Connor WE. β-sitosterolemia and xanthomatosis. A newly described lipid storage disease in two sisters. J Clin Invest 1974;53:1033–43.

[153] von Holtz RL, Fink CS, Awad AB. beta-Sitosterol activated the sphingomyelin cycle and induces apoptosis in LNCaP human prostate cancer cells. Nutr Cancer 1998;32:8–12.

[154] Patel SB, Salen G, Hidaka H, Kwiterovich PO, Stalenhoef AF, Miettinen TA, et al. Mapping a gene involved in regulating dietary cholesterol absorption. The sitosterolemia locus is found at chromosome 2p21. J Clin Invest 1998;102:1041–4.

[155] Berge KE, Tian H, Graf GA, Grishin NV, Schultz J, Kwiterovich P, et al. Accumulation of dietary cholesterol in sitosterolemia caused by mutations in adjacent ABC transporters. Science 2000;290:1771–5.

[156] Lu K, Lee MH, Hazard S, Brooks-Wilson A, Hidaka H, Kojima H, et al. Two genes that map to the STSL locus cause sitosterolemia: genomic structure and spectrum of mutations involving sterolin-1 and sterolin-2, encoded by ABCG5 and ABCG8, respectively. Am J Hum Genet 2001;69:278–90.

[157] Graf GA, Li WP, Gerard RD, Gelissen I, White A, Cohen JC, et al. Coexpression of ATP-binding cassette proteins ABCG5 and ABCG8 permits their transport to the apical surface. J Clin Invest 2002;110:659–69.

[158] Yu L, Hammer RE, Li-Hawkins J, Von Bergmann K, Lutjohann D, Cohen JC, et al. Disruption of ABCG5 and ABCG8 in mice reveals their crucial role in biliary cholesterol secretion. Proc Natl Acad Sci 2002;99:16237–42.

[159] Repa JJ, Berge KE, Pomajzl C, Richardson JA, Hobbs H, Mangelsdorf DJ. Regulation of ATP-binding cassette sterol transporters ABCG5 and ABCG8 by the liver X receptors alpha and beta. J Biol Chem 2002;277:18793–800.

[160] Escola-Gil JC, Quesada H, Julve J, Martin-Campos JM, Cedo L, Blanco-Vaca F. Sitosterolemia: diagnosis, investigation, and management. Curr Atheroscler Rep 2014;16:424.

[161] Hansel B, Carrie A, Brun-Druc N, Leclert G, Chantepie S, Coiffard AS, et al. Premature atherosclerosis is not systematic in phytosterolemic patients: severe hypercholesterolemia as a confounding factor in five subjects. Atherosclerosis 2014;234:162–8.

[162] Rubin I, Yarden Y. The basic biology of HER2. Ann Oncol 2001;12(Suppl. 1):S3–8.

[163] Dandona S. Cardiovascular drugs and the genetic response. Methodist Debakey Cardiovasc J 2014;10:13–7.

[164] Kuruvilla M, Gurk-Turner C. A review of warfarin dosing and monitoring. Proc (Bayl Univ Med Cent) 2001;14:305–6.

[165] Owen RP, Gong L, Sagreiya H, Klein TE, Altman RB. VKORC1 pharmacogenomics summary. Pharmacogenet Genom 2010;20:642–4.
[166] Rost S, Fregin A, Ivaskevicius V, Conzelmann E, Hörtnagel K, Pelz HJ, et al. Mutations in VKORC1 cause warfarin resistance and multiple coagulation factor deficiency type 2. Nature 2004;427:537–41.
[167] Yin T, Miyata T. Warfarin dose and the pharmacogenomics of CYP2C9 and VKORC1 – rationale and perspectives. Thromb Res 2007;120:1–10.
[168] Takeuchi F, McGinnis R, Bourgeois S, Barnes C, Eriksson N, Soranzo N, et al. A genome-wide association study confirms VKORC1, CYP2C9, and CYP4F2 as principal genetic determinants of warfarin dose. PLoS Genet 2009;5:e1000433.
[169] Choonara IA, Haynes BP, Cholerton S, Breckenridge AM, Park BK. Enantiomers of Warfarin and vitamin K1 metabolism. Br J Clin Pharmacol 1986;22:729–32.
[170] Breckenridge A, Orme M, Wesseling H, Lewis RJ, Gibbons R. Pharmacokinetics and pharmacodynamics of the enantiomers of warfarin in man. Clin Pharmacol Ther 1974;15:424–30.
[171] Pirmohamed M, Park BK. Cytochrome P450 enzyme polymorphisms and adverse drug reactions. Toxicology 2003;192:23–32.
[172] Kirchheiner J, Brockmöller J. Clinical consequences of cytochrome P450 2C9 polymorphisms. Clin Pharmacol Ther 2005;77:1–16.
[173] Daly AK, King BP. Contribution of CYP2C9 to variability in vitamin K antagonist metabolism. Expert Opin Drug Metab Toxicol 2006;2:3–15.
[174] McDonald MG, Rieder MJ, Nakano M, Hsia CK, Rettie AE. CYP4F2 is a vitamin K1 oxidase: an explanation for altered warfarin dose in carriers of the V433M variant. Mol Pharmacol 2009;75:1337–46.
[175] Caldwell MD, Awad T, Johnson JA, Gage BF, Falkowski M, Gardina P, et al. CYP4F2 genetic variant alters required warfarin dose. Blood 2008;111:4106–12.
[176] GGCX gamma-glutamyl carboxylase [*Homo sapiens* (human)] http://www.ncbi.nlm.nih.gov/gene/2677 [last accessed on 11.11.15].
[177] Cavallari LH, Perera M, Wadelius M, Deloukas P, Taube G, Patel SR, et al. Association of the GGCX (CAA)16/17 repeat polymorphism with higher warfarin dose requirements in African Americans. Pharmacogenet Genom 2012;22:152–8.
[178] Huang SW, Xiang DK, Huang L, Chen BL, An BQ, Li GF, et al. Influence of GGCX genotype on warfarin dose requirements in Chinese patients. Thromb Res 2011;127:131–4.
[179] Sun Y, Wu Z, Li S, Qin X, Li T, Xie L, et al. Impact of gamma-glutamyl carboxylase gene polymorphisms on warfarin dose requirement: a systematic review and meta-analysis. Thromb Res 2015;135:739–47.
[180] Rieder MJ, Reiner AP, Gage BF, Nickerson DA, Eby CS, McLeod HL, et al. Effect of VKORC1 haplotypes on transcriptional regulation and warfarin dose. N Engl J Med 2005;352:2285–93.
[181] Geisen C, Watzka M, Sittinger K, Steffens M, Daugela L, Seifried E, et al. VKORC1 haplotypes and their impact on the inter-individual and inter-ethnical variability of oral anticoagulation. Thromb Haemost 2005;94:773–9.
[182] Yuan HY, Chen JJ, Lee MT, Wung JC, Chen YF, Charng MJ, et al. A novel functional VKORC1 promoter polymorphism is associated with inter-individual and inter-ethnic differences in warfarin sensitivity. Hum Mol Genet 2005;14:1745–51.
[183] Militaru F, Vesa S, Crisan S, Militaru V, Trifa A, Buzoianu A. Genotype-phenotype correlations in patients treated with acenocoumarol. Rom Rev Lab Med 2014;22:347–54.
[184] Tatsuno S, Tatsuno E. Does ethnicity play a role in the dosing of warfarin in Hawai'i? Hawaii J Med Public Health 2014;73:76–9.

[185] Zhao F, Loke C, Rankin SC, Guo JY, Lee HS, Wu TS, et al. Novel CYP2C9 genetic variants in Asian subjects and their influence on maintenance warfarin dose. Clin Pharmacol Ther 2004;76:210–9.

[186] Voora D, Koboldt D, King C, Lenzini PA, Eby CS, Porche-Sorbet R, et al. A polymorphism in the VKORC1-regulator calumenin predicts higher warfarin doses in African-Americans. Clin Pharmacol Ther 2010;87:445–51.

[187] CYP2C9 allele nomenclature from the Human Cytochrome P450 (CYP) Allele Nomenclature Database http://www.cypalleles.ki.se/cyp2c9.htm [last accessed on 11.11.15].

[188] Van Booven D, Marsh S, McLeod H, Carillo MW, Sangkuhl K, Klein TE, et al. Cytochrome P450 2C9-CYP2C9. Pharmacogenet Genom 2010;20:277–81.

[189] Lee CR, Goldstein JA, Pieper JA. Cytochrome P450 2C9 polymorphisms: a comprehensive review of the in-vitro and human data. Pharmacogenetics 2002;12:251–63.

[190] Aithal GP, Day CP, Kesteven PJ, Daly AK. Association of polymorphisms in the cytochrome P450 CYP2C9 with warfarin dose requirement and risk of bleeding complications. Lancet 1999;353:717–9.

[191] Krishna Kumar D, Shewade DG, Loriot MA, Beaune P, Balachander J, Sai Chandran BV, et al. Effect of CYP2C9, VKORC1, CYP4F2 and GGCX genetic variants on warfarin maintenance dose and explicating a new pharmacogenetic algorithm in South Indian population. Eur J Clin Pharmacol 2014;70:47–56.

[192] Roden DM, Johnson JA, Kimmel S, Krauss RM, Wong Medina M, Shuldiner A, et al. Cardiovascular pharmacogenomics. Circ Res 2011;109:807–20.

[193] Kamali F, Wynne H. Pharmacogenetics of warfarin. Annu Rev Med 2010;61:63–75.

Chapter 10

Interactions–Interrelationships Between Genetics and Environmental Factors in Cardiovascular Disease

K. Norrington[1], E. Androulakis[2], T. Christophides[1]
[1]*Barts Health NHS Trust, London, United Kingdom;* [2]*John Radcliffe Hospital, Oxford, United Kingdom*

INTRODUCTION

Coronary heart disease (CHD) is a leading cause of mortality worldwide. Previously, disparities in rates of CHD between western nations and other countries were attributed to genetic factors. However, given that genetic predisposition is stable over many generations but the prevalence of CHD is now increasing in nonwestern countries too, the change in disease prevalence must be attributable to extrinsic lifestyle or environmental factors [1]. It is hoped that an understanding of gene–environment interaction will provide insight into the so-called missing heritability of disease mechanisms and will inform public health strategies for primary prevention of cardiovascular disease (CVD) [2]. In this chapter, specific reference is made to CHD as an archetypical example of a well-studied complex CVD.

Gene–environment interaction denotes that there is a different effect of an environmental exposure in individuals with a different genotype [1–3]. Each genotype represents a specific genetic risk profile that is modified by environmental exposure. This exposure may increase or decrease the effect of the genotype. When an interaction between the genotype and environmental factor is present, this interaction is said to exert a *main effect* on the likelihood of developing the disease, while additional independent contributions from the genotype and environmental factors are said to exert *marginal effects* only. Studies of either genetic predisposition or environmental contributing factors are often designed to only assess these marginal factors without analyzing the main effect contributed by a gene–environment interaction [4].

Disease processes such as CHD represent a disruption of physiological homeostasis [1,3] either because of intrinsic changes at, for example, the genetic level or because of exposure to extrinsic stimuli (ie, environmental factors such as smoking, diet, and diabetes). Although there are rare examples of monogenic disorders such as familial hypercholesterolemia, where specific mutations (in the low-density-lipoprotein (LDL) receptor, APOB, or PCSK9 gene) cause all individuals to exhibit the phenotype, such diseases are rare. More commonly, individuals have multiple minor mutations or genetic polymorphisms in genes coding for essential homeostatic cardiovascular processes. The cumulative effect of these individually minor polymorphisms contributes to the diversity of response to environmental stimuli between those with an otherwise identical risk profile [3]. For example, some smokers with an otherwise identical risk profile develop CHD, whereas others do not. Therefore, the interaction between these functional gene polymorphisms and environmental factors has a significant impact on CHD risk. Hence, studying these disease processes can offer significant insight into gene–environment interactions and their interrelationships.

MODELS OF INTERACTION

Interaction may be statistical or biological. It is important to distinguish interaction from confounding [4,5]. Statistical interaction does not make any inference about underlying biological causality and is purely a mathematical phenomenon whereby the level of one variable depends upon the level of a second variable such that the effects are not merely additive [4,5]. Biological interaction refers to the joint effect of two factors that act together physically in the same causal mechanism of the disease [5]. In order to quantify and draw applicable conclusions from biological interactions, measures of statistical interaction are essential and must be included in the data analysis of any gene–environment study. An essential goal of studies of gene–environment interaction is to establish new mechanisms of disease, and therefore the concept of biological interaction is crucial. Inherent to the concept of biological interaction is the concept of biological plausibility [4], which is essential for a mechanistic understanding of a proposed interaction. Confounding occurs when there is a correlation between the observed factor and a given outcome, but there is no causative relationship [5]. Therefore, an identified interaction should be examined in order to exclude confounding as a partial explanation of the observed effect.

Complementary Versus Antagonistic

Gene–environment interactions may be complementary or antagonistic [5]. With complementary interactions, genetic predisposition and environmental exposure have the same effect on the direction of disease's risk. With antagonistic interaction, the genetic predisposition and environmental exposure have opposite effects on the direction of disease risk.

Gene–Gene Interactions

The interaction between two or more genes is termed epistasis [5].

TABLE 10.1 A 2×2 Plot Demonstrating the Relative Risk (RR) of Developing a Disease in the Presence/Absence of a High-Risk Genotype in the Presence/Absence of a Given Environmental Exposure

	High-Risk Genotype Present	High-Risk Genotype Absent
Environmental exposure present	RR>>1	RR>1
Environmental exposure absent	RR>1	RR=1

Gene–environment correlations are distinct from gene–environment interactions [5]. Correlations occur when genetic markers are highly associated with and possibly cause a given behavioral characteristic or environmental exposure.

Models of Disease Risk

The classic 2×2 plot is typically used to elucidate the risk of a disease in the presence of environmental and genetic risk factors [5]. This simplified model presented by the 2×2 plot assumes that the disease's outcome is dichotomous [5], that the gene–environment interaction is complementary, and that the relative risk (RR) of the general population is 1 (Table 10.1).

Additive

The effect of each predictor must be summed in order to determine the likelihood (or relative risk (RR)) of disease in a given subject, taking into account both genetic and environmental factors (RR_{GE}) relative to the general population, such that:

$$RR_{GE} = (RR_g + RR_e) - 1$$

The relative risk of a given genotype (RR_g) and environmental exposure (RR_e) may be complementary or antagonistic. However, when considering gene–environment interaction, it is important to emphasize that in the cases of additive models, the net effect of all predictors is the sum of each predictor's individual effect on the disease outcome, whereas in gene–environment interactions the effect of the environment is modified by the subject's genotype [4,5].

Multiplicative

In a multiplicative model, the effect of the genotype of the risk profile of the individual concerned depends upon environmental exposure [5].

$$RR_{GE} = RR_g \times RR_e - 1$$

Additive Versus Multiplicative Model

The choice of model appropriate for a given study depends upon the biological relationship between the predictors of risk. For example, if the predictors act upon the same pathway, then an additive model is assumed, whereas if predictors act upon two separate pathways, a multiplicative model is assumed. Second, the aim of the study is important. For example, in public health models, an additive model may suffice because the predictive capability of the model is paramount, whereas in studies of disease etiology where the intention is to understand mechanistic processes, a multiplicative process may be necessary to reflect the underlying biology. Frequently, statistical modeling of both additive and multiplicative types is performed in order to determine the possible implications of incorrect model application. If the models do not differ, the additive model is often chosen given the greater simplicity [4,5]. Prospective population-based cohorts are the gold standards for gene–environment interaction studies because they reduce the change of measurement error and therefore subsequent bias; however, they are time-consuming and resource intensive [4].

ENVIRONMENTAL FACTORS

A risk factor is defined as a measurable characteristic that is causally associated with the development of diseases. The principle and importance of risk factors in disease development were first established in the pioneering epidemiological Framingham Heart Study, which established hypertension, hyperlipidemia, and significant causal factors in the development of atherosclerotic coronary artery disease (CAD). In addition to diabetes and smoking, hypertension and hyperlipidemia represent the four conventional risk factors for CAD, which are present in over 80% of people who develop CAD, with smoking significantly reducing the age at first presentation with a cardiac event [6]. However, individuals with these risk factors do not always develop the disease to the same extent. Also, novel risk factors have been recently proposed to be associated with CAD (Table 10.2). Additional environmental exposures such as aerosolized pollutants are now established as agents of increased CVD morbidity. Proposed gene–environment interactions to explain these differences will also be discussed.

INTERACTIONS BETWEEN GENOTYPE, DIET, AND CORONARY HEART DISEASE

Diet plays an important role in the development of CHD. Diets high in trans fatty acids, with high glycemic load, and low in raw vegetables, fruit, nuts, and whole grains are associated with myocardial infarction (MI) in cohort studies [7].

Genome-wide association studies have identified genetic variants of regions of the chromosome 9p21 region that are significantly associated with cardiovascular disease. One study genotyped four known risk alleles in the 9p21 region associated with increasing risk of cardiovascular disease in over 8000

TABLE 10.2 Classic and Novel Risk Factors for Atherosclerosis

	Classic	Novel
Gender	Prehypertension	Lipoprotein(a)
Age	Metabolic syndrome	Homocysteine
Family history	Diabetes mellitus	Fibrinogen
Hypertension	Impaired glucose tolerance	C-reactive protein
Smoking	Insulin resistance	Other inflammatory biomarkers
Cholesterol/LDL	Hypertriglyceridemia	Infectious agent

LDL, Low-density lipoprotein.

European, South Asian, Chinese, and Latin American patients enrolled in the INTERHEART study. The effect of the risk alleles located in the 9p21 region on MI was reduced in patients with a diet rich in raw vegetables and fruits. These findings were corroborated in a cohort of over 19,000 patients in the FINRISK study [8]. Stricter adherence to the Mediterranean-type diet pattern has been shown to counteract the genetic predisposition to develop type 2 diabetes posed by the FTO and MC4R polymorphisms [9].

The results of multiple studies into the contribution of coffee and intake to cardiovascular disease have been contradictory, and a complete understanding of its role remains unclear. However, the main active ingredient in coffee, caffeine, is metabolized by CYP1A2, which displays common interindividual variation whereby a common genetic polymorphism, CYP1A2*1F, results in a reduced rate of caffeine metabolizing. One case-controlled study examined the association of coffee intake with relative risk of nonfatal MI, and it was found that the intake of coffee was associated with an increased risk of nonfatal MI only among individuals with slow caffeine metabolism [10]. Potential mechanisms include reduced enzyme levels in individuals with the genetic polymorphism. This and other such studies are unable to demonstrate causality, and are limited by patient-reported consumption questionnaires. Furthermore, coffee has a number of constituent compounds, and distinguishing between the effects of caffeine and those other compounds is difficult.

INTERACTIONS BETWEEN ALCOHOL AND CORONARY HEART DISEASE

Moderate alcohol consumption is consistently associated with reduced risk of acute coronary events such that there is a well-established inverse J-shaped curve between alcohol and CVD morbidity for which the most plausible biological mechanism is thought to be through increase in high-density-lipoprotein

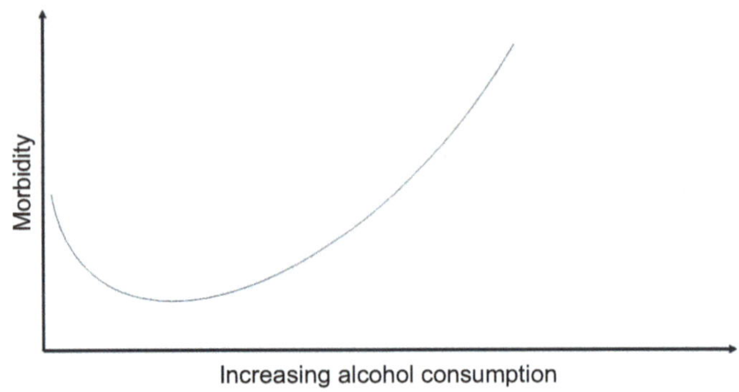

FIGURE 10.1 J-shaped curve illustrating the relationship between alcohol consumption and cardiovascular disease risk.

(HDL) cholesterol [11] (Fig. 10.1). Alcohol is metabolized by alcohol dehydrogenase, which exists in multiple isoforms: *ADH1*, *ADH2*, and *ADH3*. Hines and colleagues examined the ADH3 genotype among men in the Physicians Health Study and controls. Homozygosity for the allele associated with a slow rate of ethanol oxidation (gamma2) was associated with a reduced risk of MI when compared to homozygosity for the allele associated with a fast rate of ethanol oxidation (gamma1). Most interestingly, moderate alcohol consumption was associated with a decreased risk of MI in all three genotype groups (gamma1gamma1, gamma1gamma2, or gamma2gamma2), with the greatest reduction in risk seen in men with moderate alcohol consumption who are homozygous for the gamma2 allele. These men also had higher HDL levels [12].

INTERACTION BETWEEN APOE GENOTYPE, SMOKING, AND RISK OF CORONARY HEART DISEASE

Apolipoprotein E (apoE) is an intermediate-density lipoprotein that is responsible for forming and exporting a stabilized triacylglycerol and cholesterol lipid core from the liver into the blood for transport to fatty acid–utilizing tissues. Genetic variation in the common apoE isoforms E2, E3, and E4 affects plasma lipid levels, and apoE was one of the first genetic polymorphisms considered to play a role in the pathogenesis of atherosclerosis and therefore the risk of CHD.

A number of population-based studies have attempted to assess the impact of the APOE genotype, smoking, and CHD risk: the Northwick Park Heart Study, the Framingham Offspring Study (FOS), and National Heart, Lung, and Blood Institute Family Heart Study, which will be briefly reviewed. Smoking is a well-established environmental risk factor for CAD, and therefore there has been significant investigation as to whether there is an interaction between smoking, the APOE genotype, and CAD.

The prospective second Northwick Park Heart Study enrolled 3052 men aged 50–61 years who were free of any history of ischemic heart disease.

Participants were divided into smokers, ex-smokers, and lifelong nonsmokers. Smoking habit was established by a questionnaire at baseline and at annual follow-up review. APOE genotype was established using polymerase chain reaction (PCR) analysis. Over the follow-up period, an ex-smoker was 1.34 times more likely and a smoker 1.94 times more likely to develop CAD (fatal CHD, nonfatal MI, and coronary artery surgery and silent MI) compared to nonsmokers, respectively. In never-smokers, risk was roughly the same irrespective of genotype. When stratified according to the apoE allele, smokers who were carriers of the E4 allele had a significantly increased risk of CAD compared with the nonsmoking group [13]. It should be noted that these findings remained significant when adjustment for lipid variables was made. These findings provide strong evidence for the context dependency of APOE genotype risk, that is, the phenotypic outcome depends not just upon the genotype but also upon the environmental context (smoking) to which a given genotype (in this case, isoforms of the APOE genotype) is exposed.

In the prospective Framingham Offspring Study, the interaction between the APOE genotype and CVD was investigated in over 3400 study participants. After adjustment for nonlipid risk factors, including smoking in men, the odds ratio for cardiovascular disease was 1.79 and 1.62 for E2 and E4 men, respectively. The differential risk was attributed by the authors to the presence of apoE allele (epsilon) 2 or apoE allele (epsilon) 4 [14]. However, subsequent reanalysis with the specific intention of examining the APOE genotype–smoking interaction demonstrated that the significantly increased CAD risk was only present in E4 men who smoked. These findings have been corroborated by the National Heart, Lung, and Blood Institute Family Heart Study, which showed a synergistic effect of the apoE allele (epsilon) 4, smoking, and ultrasound-diagnosed carotid artery atherosclerosis, which is often used as a general marker of generalized atherosclerosis [15]. LDL oxidation is thought to be a critical step in the putative mechanism uniting the apoE allele (epsilon) 4, smoking interaction, and CHD risk [1].

ENVIRONMENTAL POLLUTANTS, EPIGENETICS, AND CARDIOVASCULAR DISEASE

Ambient air contains a mixture of variable-density aerosolized particles or particulate matter (PM), liquids, and gases. The majority of air pollution in developed nations is derived from the combustion of fossil fuels. Despite the intuitive assumption that the main effect of PM would be on respiratory disease, the majority of evidence-based adverse effects are on the cardiovascular system [16] (Fig. 10.2). Increases in the concentration of PM are associated with increased mortality and hospital attendance. There are statistically significant associations of fluxes in PM with ischemic heart disease, arrhythmias, heart failure, and cardiac arrest. Significant further cross-disciplinary work is required in order to develop greater mechanistic insight. Fluxes in various atmospheric gases,

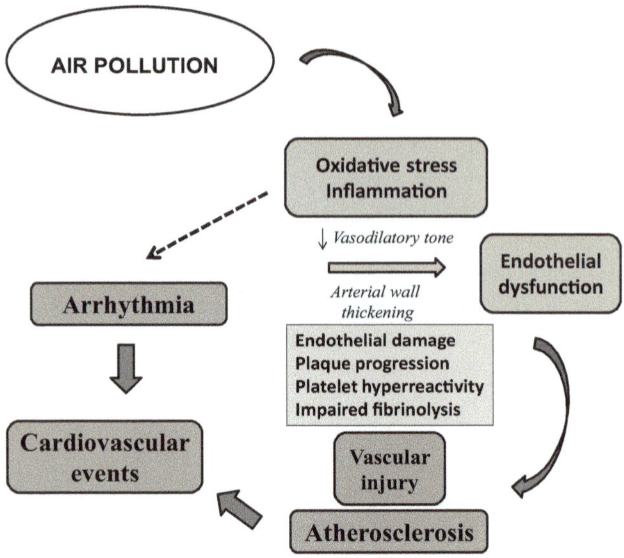

FIGURE 10.2 Proposed mechanisms connecting air pollution and cardiovascular disease.

including CO, NO_2, SO_2, and ozone, have been linked to cardiovascular events, including acute coronary events [17] (ozone), hospitalization for heart failure [18] (CO, NO_2, and SO_2), and arrhythmias [19].

Although these associated have been corroborated in long and short-term exposure studies, the mechanisms remain unclear. There has been no satisfactory explanation for how remote inhalation of PM into the lungs is capable of affecting distant cardiovascular territories, and theories have typically centered on how PM drives autonomic dysregulation, the release of pro-inflammatory and pro-oxidative mediators from the lungs, and nanoscale soluble particles translocating from the lungs into the systemic circulation and then interacting directly with the cardiovascular system [20].

It is becoming increasingly apparent that epigenetics may play an important role in environmental pollutant–driven CVD (Fig. 10.3). Epigenetics allows genes to be silenced at different points during an organism's life cycle. DNA methylation is one of several epigenetic mechanisms used to control gene expression and lock genes in the "off" position [21]. In animal studies, it has been shown that inhibition of the long interspersed nuclear element 1 (LINE-1) activity in the heart decreases ischemic damage [22]. One study evaluated whether particulate pollution modifies DNA methylation of LINE-1 in 718 elderly patients in the Boston Area Normative study. Demethylation decreases rapidly following peaks of ambient levels of PM, thus providing a putative mechanism for environmentally driven changes in ischemic events [23]. However, these results must be interpreted with caution as they cannot demonstrate causality but do provide a promising avenue for further investigation.

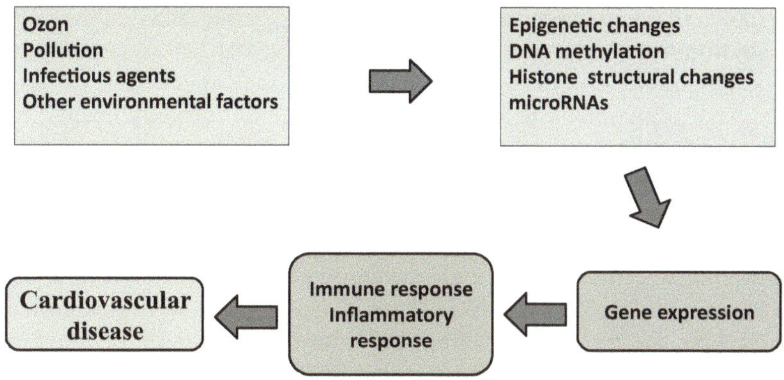

FIGURE 10.3 Interrelationships among environmental factors, epigenetics, and cardiovascular disease.

DRUG–GENOME INTERACTIONS RELEVANT TO CARDIOVASCULAR DISEASE

Warfarin is a vitamin K antagonist that is important in the prevention of thromboembolic events such as thromboembolic stroke secondary to atrial fibrillation. In individuals with appropriate risk factors, the efficacy of warfarin depends upon the amount of time that an individual spends with an international normalized ratio (INR) in the therapeutic range, with risk of severe bleeding (eg, intracranial hemorrhage) if the INR is too high and risk of thromboembolic events if it is too low. However, widespread interindividual variation in the metabolism of warfarin means that establishing the appropriate dosage regime to achieve sufficient time in the therapeutic range can be difficult to achieve in the clinical setting.

Evidence from genome-wide association studies has shown that genotype contributes to interindividual variability of response to warfarin therapy. These studies have identified *CYP2C9* and *VKORC1* as critical in the pharmacokinetics of warfarin metabolism [24]. Of the 35 known *CYP2C9* alleles known, the CYP2C9*2 and *3 single nucleotide polymorphisms (SNPs) are the best studied. Warfarin clearance has been shown to be reduced by approximately 40% in patients with the *CYP2C9*2* polymorphisms and by 75% with *CYP2C9*3* [25]. Patients with these polymorphisms must have significantly different dosage requirements in order to achieve the same therapeutic INR [26]. *VKOR1* encodes the drug target of warfarin: the vitamin K epoxide reductase enzyme. A common polymorphism of the *VKOR1* is −1639G>A, which occurs in a promoter region of the gene and alters transcription factor binding, with carriers of −1639A requiring lower initial warfarin dosing than −1639G carriers [27,28]. These studies have demonstrated different allele frequencies among different racial groups, which suggest an explanation for observed differences in warfarin dosing requirements. Pharmacogenetics-based warfarin dosing algorithms have been shown to predict appropriate warfarin dosing better than clinical or fixed-dose algorithms [29].

Statins are HMG–CoA reductase inhibitors that lower LDL cholesterol levels; have been shown to reduce the incidence of coronary events, revascularization, and stroke; and are therefore essential drugs in the prevention of major adverse cardiac events in patients at high risk for CAD [30]. However, treatment nonadherence occurs in up to 26.6% of patients on statin therapy and has been associated with worse outcomes in these patients [31]. Statin-induced side effects include myalgia, creatine kinase elevations, and rhabdomyolysis. In the STRENGTH study, participants were randomized to therapy with one of the following: atorvastatin, simvastatin, or pravastatin. With regard to an SNP of the cytochrome P450 enzyme, SLCO1B1*5 genotype and female sex were associated with statin-induced side effects that were markedly reduced when treated with pravastatin [32]. These results have potential implications for clinical practice, where, based upon existing data, pravastatin could be a preferred first choice for carriers of the SLCO1B1*5 genotype.

GENE–ENVIRONMENTAL INTERACTIONS AND CONGENITAL HEART DISEASE

Congenital heart defects remain the most common birth defects, and they therefore represent an important public health problem. Adequate intake of folic acid in the time period around conception has been demonstrated to be protective against congenital defects, although the metabolic pathway is unclear. However, it is thought to be multifactorial with a common endpoint of altered folate metabolism. It has therefore been proposed that maternal genetic polymorphisms in pathways responsible for folate metabolism may contribute to the development of congenital heart disease. One such SNP is the 677C>T (rs1801133) in the methylenetetrahydrofolate reductase gene (MTHFR 677C>T). One study demonstrated that obese women carrying the MTHFR TT genotype were 4.6 times more likely to have an affected pregnancy compared to those who were not [33]. This showed that functional polymorphisms in the folate metabolic pathway are associated with an increased risk of having a child with congenital heart disease in mothers who are obese, smoke or drink alcohol, and possess this genetic polymorphism.

CONCLUSIONS

With only a few noted exceptions, CVD increasingly shows that the environmental context in which a genotype is expressed may be critical to the risk of disease development; a genetic polymorphism may have a minor effect in individuals who maintain a low environmental risk but have a major effect in a "high-risk" environment. Knowledge of these gene–environment interactions provides mechanistic insight into observed differences of disease risk in individuals with an otherwise identical risk profile (eg, APOE genotype, smoking, and risk of CHD). In an era of evidence-based medicine where there is an ever-expanding arsenal of prognostically beneficial drugs, pharmacogenetics is a rapidly developing field that has the potential to add a personalized and nuanced dimension to patient care. However, much prospective work is required to translate data from large-scale clinical trials to the patient at the bedside.

REFERENCES

[1] Talmud PJ. Gene-environment interaction and its impact on coronary heart disease risk. Nutr Metab Cardiovasc Dis 2007;17:148–52.

[2] Hunter DJ. Gene-environment interactions in human diseases. Nat Rev Genet 2005;6:287–98.

[3] Talmud PJ. How to identify gene-environment interactions in a multifactorial disease: CHD as an example. Proc Nutr Soc 2004;63:5–10.

[4] Ottman R. An epidemiologic approach to gene-environment interaction. Genet Epidemiol 1990;7:177–85.

[5] Flowers E, Froelicher ES, Aouizerat BE. Gene-environment interactions in cardiovascular disease. Eur J Cardiovasc Nurs 2012;11:472–8.

[6] Khot UN, et al. Prevalence of conventional risk factors in patients with coronary heart disease. JAMA 2003;290:898–904.

[7] Mente A, de Koning L, Shannon HS, Anand SS. A systematic review of the evidence supporting a causal link between dietary factors and coronary heart disease. Arch Intern Med 2009;169:659–69.

[8] Do R, et al. The effect of chromosome 9p21 variants on cardiovascular disease may be modified by dietary intake: evidence from a case/control and a prospective study. PLoS Med 2011;8.

[9] Liu C, Mou S, Pan C. The FTO gene rs9939609 polymorphism predicts risk of cardiovascular disease—a systematic review and meta-analysis. PLoS One 2013;8:e71901.

[10] El-Sohemy A, Cornelis MC, Kabagambe EK, Campos H. Coffee, CYP1A2 genotype and risk of myocardial infarction. Genes Nutr 2007;2:155–6.

[11] Sesso HD, Gaziano JM. Alcohol intake and cardiovascular morbidity and mortality. Curr Opin Nephrol Hypertens 1999;8:353–7.

[12] Hines LM, et al. Genetic variation in alcohol dehydrogenase and the beneficial effect of moderate alcohol consumption on myocardial infarction. N Engl J Med 2001;344:549–55.

[13] Humphries SE, et al. Apolipoprotein E4 and coronary heart disease in middle-aged men who smoke: a prospective study. Lancet 2001;358:115–9.

[14] Lahoz C, et al. Apolipoprotein E genotype and cardiovascular disease in the Framingham heart study. Atherosclerosis 2001;154:529–37.

[15] Djoussé L, et al. Smoking influences the association between apolipoprotein E and lipids: the national heart, lung, and blood Institute family heart study. Lipids 2000;35:827–31.

[16] Bhatnagar A. Environmental cardiology: studying mechanistic links between pollution and heart disease. Circ Res 2006;99:692–705.

[17] Ruidavets J-B, et al. Ozone air pollution is associated with acute myocardial infarction. Circulation 2005;111:563–9.

[18] Wellenius GA, Bateson TF, Mittleman MA, Schwartz J. Particulate air pollution and the rate of hospitalization for congestive heart failure among medicare beneficiaries in Pittsburgh, Pennsylvania. Am J Epidemiol 2005;161:1030–6.

[19] Routledge HC, Manney S, Harrison RM, Ayres JG, Townend JN. Effect of inhaled sulphur dioxide and carbon particles on heart rate variability and markers of inflammation and coagulation in human subjects. Heart 2006;92:220–7.
[20] Brook RD. Cardiovascular effects of air pollution. Clin Sci (Lond) 2008;115:175–87.
[21] Phillips T. The role of methylation in gene expression. Nat Educ 2008;1(1).
[22] Lucchinetti E, et al. Inhibition of LINE-1 expression in the heart decreases ischemic damage by activation of Akt/PKB signaling. Physiol Genom 2006;25:314–24.
[23] Baccarelli A, et al. Rapid DNA methylation changes after exposure to traffic particles. Am J Respir Crit Care Med 2009;179:572–8.
[24] Takeuchi F, et al. A genome-wide association study confirms VKORC1, CYP2C9, and CYP4F2 as principal genetic determinants of warfarin dose. PLoS Genet 2009;5.
[25] Scordo MG, et al. Influence of CYP2C9 and CYP2C19 genetic polymorphisms on warfarin maintenance dose and metabolic clearance. Clin Pharmacol Ther 2002;72:702–10.
[26] Lindh JD, Holm L, Andersson ML, Rane A. Influence of CYP2C9 genotype on warfarin dose requirements-a systematic review and meta-analysis. Eur J Clin Pharmacol 2009;65:365–75.
[27] Rieder MJ, et al. Effect of VKORC1 haplotypes on transcriptional regulation and warfarin dose. N Engl J Med 2005;352:2285–93.
[28] Yuan HY, et al. A novel functional VKORC1 promoter polymorphism is associated with inter-individual and inter-ethnic differences in warfarin sensitivity. Hum Mol Genet 2005;14: 1745–51.
[29] Klein TE, et al. Estimation of the warfarin dose with clinical and pharmacogenetic data. N Engl J Med 2009;360:753–64.
[30] Cholesterol Treatment Trialists' (CTT) Collaborators. Efficacy of cholesterol-lowering therapy in 18,686 people with diabetes in 14 randomised trials of statins: a meta-analysis. Lancet 2008;371:117–25.
[31] Ho PM, et al. Medication nonadherence is associated with a broad range of adverse outcomes in patients with coronary artery disease. Am Heart J 2008;155:772–9.
[32] Voora D, et al. The SLCO1B1*5 genetic variant is associated with statin-induced side effects. J Am Coll Cardiol 2009;54:1609–16.
[33] Hobbs CA, Cleves MA, Karim MA, Zhao W, MacLeod SL. Maternal folate-related gene environment interactions and congenital heart defects. Obstet Gynecol 2010;116:316–22.

Index

'*Note*: Page numbers followed by "f" indicate figures and "t" indicate tables.'

D

H

I

J

L

M

N

O

Kolibri 360 GmbH
Ferdinand-Jühlke-Straße 7
99095 Erfurt, Deutschland
produktsicherheit@kolibri360.de